Introduction to
Hydrodynamic Stability

Instability of flows and their transition to turbulence are widespread phenomena in engineering and the natural environment, and are important in applied mathematics, astrophysics, biology, geophysics, meteorology, oceanography and physics as well as engineering. This is a textbook to introduce these phenomena at a level suitable for a graduate course, by modelling them mathematically, and describing numerical simulations and laboratory experiments. The visualization of instabilities is emphasized, with many figures, and in references to more still and moving pictures. The relation of chaos to transition is discussed at length. Many worked examples and exercises for students illustrate the ideas of the text. Readers are assumed to be fluent in linear algebra, advanced calculus, elementary theory of ordinary differential equations, complex variables and the elements of fluid mechanics. The book is aimed at graduate students but will also be very useful for specialists in other fields.

Philip Drazin (1934–2002) was Professor of Applied Mathematics at the University of Bristol (1981–1999) and Professor of Mathematical Sciences at the University of Bath (1999–2002). He was the author of many books including *Hydrodynamic Stability* which he co-authored with W. H. Reid (Cambridge University Press, 1981).

Cambridge Texts in Applied Mathematics

Thinking About Ordinary Differential Equations
Robert E. O'Malley

A Modern Introduction to the Mathematical Theory of Water Waves
R.S. Johnson

Rarefied Gas Dynamics
Carlo Cercignani

Symmetry Methods for Differential Equations
Peter E. Hydon

High Speed Flow
C.J. Chapman

Wave Motion
J. Billingham and A.C. King

An Introduction to Magnetohydrodynamics
P.A. Davidson

Linear Elastic Waves
John G. Harris

Vorticity and Incompressible Flow
Andrew J. Majda and Andrea L. Bertozzi

Infinite Dimensional Dynamical Systems
James C. Robinson

An Introduction to Symmetry Analysis
Brian J. Cantwell

Bäcklund and Darboux Transformations
C. Rogers and W.K. Schief

Finite-Volume Methods for Hyperbolic Problems
Randall J. LeVeque

From the Editors

We are deeply saddened to note the death of Philip Drazin during the production of this, his last textbook for the Cambridge Texts in Applied Mathematics. Philip was a wonderful teacher, a superb applied mathematician and an inspiring colleague. During his life he produced seminal work on hydrodynamic stability, particularly applied to problems in meteorology. He was always concerned to understand the mathematics behind the physical problem he was studying, but was always aware of its limitations, and the need to compare mathematical predictions against physical reality.

This book is a fitting tribute to Philip's whole approach to his work. It reflects both his deep understanding of the way mathematics can be applied to natural phenomena and his unique way of illuminating any topic. All who knew him will see his spirit and humour shining through these pages and will benefit from the experience and wisdom he gained by studying many significant practical and theoretical problems. Philip wrote two earlier textbooks within this series, each going on to become classics in their fields. We are sure that this will do the same.

Introduction to Hydrodynamic Stability

P. G. DRAZIN

CAMBRIDGE
UNIVERSITY PRESS

32 Avenue of the Americas, New York NY 10013-2473, USA

Cambridge University Press is part of the University of Cambridge.

It furthers the University's mission by disseminating knowledge in the pursuit of education, learning and research at the highest international levels of excellence.

www.cambridge.org
Information on this title: www.cambridge.org/9780521009652

© Cambridge University Press 2002

First published 2002

A catalogue record for this publication is available from the British Library

Library of Congress Cataloguing in Publication data

Drazin, P. G.
　　Introduction to hydrodynamic stability / P. G. Drazin.
　　　　p.　cm. – (Cambridge texts in applied mathematics; 32)
　　Includes bibliographical references and index.
　　ISBN 0-521-80427-2 – ISBN 0-521-00965-0 (pb.)
　　　1. Hydrodynamics.　2. Stability.　I. Title.　II. Series.
　　QA911.D73　　2002
　　532′.5 – dc21　　　　　　　　　　　　　　　　2002025690

ISBN 978-0-521-80427-1 Hardback
ISBN 978-0-521-00965-2 Paperback

To Judith

Contents

xi

Preface

This text arose from notes on lectures delivered to M.Sc. students at the University of Bristol in the 1980s. The notes were revised and printed for a course of lectures delivered to postgraduates at the University of Tokyo in 1995. The latter course led to collaboration with Professor Tsutomu Kambe in writing in Japanese the book *Ryutai Rikigaku – Anteisei To Ranyu (Fluid Dynamics – Stability and Turbulence)*, published by the University of Tokyo Press in 1998. The present book is an enlargement in English of the first part of the Japanese book. An advanced draft was prepared for a lecture course given to undergraduates and postgraduates at the University of Oxford in 2001. I am grateful to the many students, at Bristol, Tokyo and Oxford, for their stimulating me to clarify both my ideas and their expression, and their encouragement to learn more. I am especially grateful to Professor Kambe for what I learnt from him and put into the text.

The result is a textbook, not a research monograph. To be sure, many points of current research have been incorporated in the text, but there has been no attempt to lead the reader up to the frontier of current research. So the mathematical theory has been described as simply and briefly as was felt possible, and plenty of worked examples and accessible exercises for students have been included. I have cited many publications, perhaps because the habit of doing so is deeply ingrained, certainly not because I ever imagined that many students care about references, let alone follow them up. The overt intention of including the references is to encourage students' instructors to follow up various details and, most importantly, use still and moving pictures to supplement this book in their teaching.

Indeed, wherever practical, pictures of relevant fluid mechanical experiments are used in the text. This is done primarily by inclusion of illustrations in this book. However, practical limitations of space have led to supplementation of the illustrations in this book by citing other sources, notably the beautiful

xv

books *An Album of Fluid Motion*, edited by Van Dyke (1982), and *Visualized Flow*, edited by Nakayama (1988). But hydrodynamic instability is essentially dynamic, so motion pictures and videos can convey many things which still pictures cannot. Accordingly, reference is often made to the wonderful classic series of film loops and motion pictures of the National Education Center; they are old and no longer for sale, but they have been re-issued as videos by the Encyclopedia Britannica Corporation. Further, *Multi-media Fluid Mechanics*, a compact disk by Homsy *et al.* (CD2000), has been published recently. Its section *Video Library* has many short videos relevant to this book, and they are cited in the text. I hope that further videos will be added to the CD in future editions, and am confident that advances in computer technology will soon lead to more such pictorial aids to this book.

It is assumed that readers of this book are familiar with the elements of the theory and practice of fluid mechanics – the material that is included in typical first courses on the motion of inviscid and viscous fluids. So the theory of Euler's equations of motion, irrotational flow, vorticity, the Navier–Stokes equations, boundary-layer theory, separation, and so forth will be used with little explanation wherever they are needed in the text. Again, the elementary theory of linear algebra, complex variables, and ordinary and partial differential equations will be assumed, and used freely as needed. Sections, paragraphs and exercises that demand more advanced knowledge or touch deep matters are preceded by asterisks.

I thank Professor William H. Reid for his generosity in allowing me to reproduce with little alteration Sections 1, 4 and 5 of our book *Hydrodynamic Stability* as respectively Section 1.1, Chapter 3 and Chapter 4 of this book, as well as several exercises. I also thank him for the enormous amount about writing books as well as about hydrodynamic stability which I have learnt from him over many decades.

I thank Professor Herbert E. Huppert, Dr Richard R. Kerswell and Professor Stephen D. Mobbs for suggesting ideas which have led to exercises in this book.

I thank Dr Alan McAlpine for material for Figure 8.10.

I thank Dr Álvaro Meseguer and Professor L. Nicholas Trefethen for their constructive comments on parts of a draft of the book, and for copies of Figures 8.14 and 8.15.

I thank Professor William Saric for an illuminating discussion of the flow past a flat plate.

I thank the reproduction-rights holders for their generous permission to reproduce many of the figures in this book, and to the authors who have kindly expressed their approval of the reproduction of their original figures.

Philip Drazin
University of Bath
July 2001

1

General Introduction

Whosoever loveth instruction loveth knowledge. . . .

Prov. xii 1

In this chapter the text begins with an informal introduction to the concept of stability and the nature of instability of a particular flow as a prototype – the flow along a pipe. The prototype illustrates the importance of instability as a prelude to transition to turbulence. Finally, the chief methods of studying instability of flows are briefly introduced.

1.1 Prelude

Hydrodynamic stability concerns the stability and instability of motions of fluids.

The concept of stability of a state of a physical or mathematical system was understood in the eighteenth century, and Clerk Maxwell (see Campbell & Garnett, 1882, p. 440) expressed the qualitative concept clearly in the nineteenth:

When . . . an infinitely small variation of the present state will alter only by an infinitely small quantity the state at some future time, the condition of the system, whether at rest or in motion, is said to be stable; but when an infinitely small variation in the present state may bring about a finite difference in the state of the system in a finite time, the condition of the system is said to be unstable.

So hydrodynamic stability is an important part of fluid mechanics, because an unstable flow is not observable, an unstable flow being in practice broken down rapidly by some 'small variation' or another. Also unstable flows often evolve into an important state of motion called *turbulence*, with a chaotic three-dimensional vorticity field with a broad spectrum of small temporal and spatial scales called *turbulence*.

The essential problems of hydrodynamic stability were recognized and formulated in the nineteenth century, notably by Helmholtz, Kelvin, Rayleigh and Reynolds. It is difficult to introduce these problems more clearly than in Osborne Reynolds's (1883) own description of his classic series of experiments on the instability of flow in a pipe, that is to say, a tube (see Figure 1.1 for

1

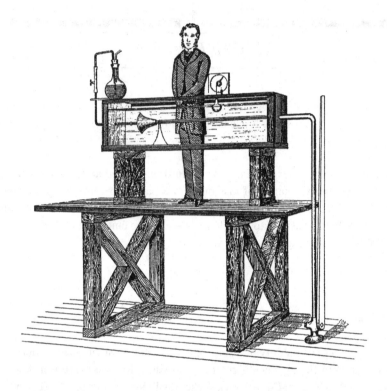

Figure 1.1   The configuration of Reynolds's experiment on flow along a pipe. (From Reynolds, 1883, Fig. 13.)

the general configuration of his apparatus, with an unnamed Victorian man to scale it).

The ... experiments were made on three tubes .... The diameters of these were nearly 1 inch, $\frac{1}{2}$ inch and $\frac{1}{4}$ inch. They were all ... fitted with trumpet mouthpieces, so that the water might enter without disturbance. The water was drawn through the tubes out of a large glass tank, in which the tubes were immersed, arrangements being made so that a streak or streaks of highly coloured water entered the tubes with the clear water.

The general results were as follows:–

(1) When the velocities were sufficiently low, the streak of colour extended in a beautiful straight line through the tube, Figure 1.2(a).

(2) If the water in the tank had not quite settled to rest, at sufficiently low velocities, the streak would shift about the tube, but there was no appearance of sinuosity.

(3) As the velocity was increased by small stages, at some point in the tube, always at a considerable distance from the trumpet or intake, the colour band would all at once mix up with the surrounding water, and fill the rest of the tube with a mass of coloured water, as in Figure 1.2(b).

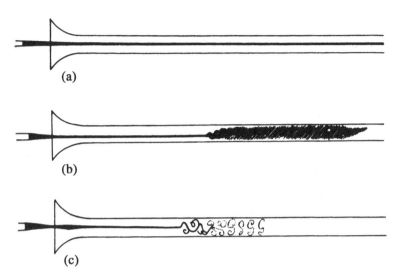

Figure 1.2   Sketches of (a) laminar flow in a pipe, indicated by a dye streak; (b) transition to turbulent flow in a pipe; and (c) transition to turbulent flow as seen when illuminated by a spark. (From Reynolds, 1883, Figs. 3, 4 and 5.)

Any increase in the velocity caused the point of break down to approach the trumpet, but with no velocities that were tried did it reach this.

On viewing the tube by the light of an electric spark, the mass of colour resolved itself into a mass of more or less distinct curls, showing eddies, as in Figure 1.2(c).

Reynolds went on to show that the *laminar flow*, the smooth flow he described in paragraph (1), breaks down when Va/v exceeds a certain critical value, V being the maximum velocity of the water in the pipe, a the radius of the pipe, and v the kinematic viscosity of water at the appropriate temperature. This dimensionless number Va/v, now called the *Reynolds number*, specifies any class of dynamically similar flows through a pipe; here we shall denote the number by R. The series of experiments gave the critical value R_c of the Reynolds number as nearly 13 000. However,

the critical velocity was very sensitive to disturbance in the water before entering the tubes . . . .

This at once suggested the idea that the condition might be one of instability for disturbance of a certain magnitude and [stability] for smaller disturbances.

Just above the critical velocity

Another phenomenon . . . was the intermittent character of the disturbance. The disturbance would suddenly come on through a certain length of the tube and pass away

Figure 1.3   Crude sketch of turbulent spots in a pipe. (From Reynolds, 1883, Fig. 16.)

and then come on again, giving the appearance of flashes, and these flashes would often commence successively at one point in the pipe. The appearance when the flashes succeeded each other rapidly was as shown in Figure 1.3.

Such 'flashes' are now called *turbulent spots* or *turbulent bursts*. Below the critical value of the Reynolds number there was laminar Poiseuille pipe flow with a parabolic velocity profile, the resistance of the pipe (that is, the tube) to the flow of water being proportional to the mean velocity. As the velocity increased above its critical value, Reynolds found that the flow became *turbulent*, with a chaotic three-dimensional motion that strongly diffused the dye throughout the water in the pipe. The resistance of the pipe to turbulent flow grew in proportion to the square of the mean velocity.

Reynolds's original apparatus survives in Manchester in England, and was used in the 1970s to repeat his experiment. You can therefore see (Van Dyke, 1982, Fig. 103) photographs of the flow in Reynolds's apparatus.

Later experimentalists have introduced perturbations, that is to say, disturbances, of finite amplitude at the intake or used pipes with roughened walls to find R_c as low as 2000, and have used such regular flows and such smooth-walled pipes that R_c was 10^5 or even more. Reynolds's description illustrates the aims of the study of hydrodynamic stability: to find whether a given laminar flow is unstable and, if so, to find how it breaks down into turbulence or some other laminar flow.

Methods of analysing the stability of flows were formulated in Reynolds's time. The method of normal modes for studying the oscillations and instability of a dynamical system of particles and rigid bodies was already highly developed. A known solution of Newton's or Lagrange's equations of motion for the system was perturbed. The equations were linearized by neglecting products of the perturbations. It was further assumed that the perturbation of each quantity could be resolved into independent components or modes varying with time t like e^{st} for some constant s, which is in general complex. The values of s for the modes were calculated from the linearized equations. If the real part of s was found to be positive for any mode, the system was deemed unstable because a general initial small perturbation of the system would grow exponentially in time until it was no longer small. Stokes, Kelvin and Rayleigh adapted this method of normal modes to fluid dynamics. An essential mathematical difference between fluid and particle dynamics is that the equations of motion are partial rather than ordinary differential equations. This difference leads to many technical difficulties in hydrodynamic stability, which, to this day, have been fully overcome for only a few classes of flows with simple configurations.

Indeed, Reynolds's experiment itself is still imperfectly understood (Eliahou *et al.*, 1998). However, we can explain qualitatively the transition from laminar flow to turbulence with some confidence. Poiseuille pipe flow with a parabolic profile is stable to infinitesimal perturbations at all Reynolds numbers. At sufficiently small values of the Reynolds number, for $R \leq R_g$, say, all perturbations, large as well as small, of the parabolic flow decay eventually; observation shows that $R_g \approx 2000$. Some way below the observed critical Reynolds number, a perturbation may grow if it is not too small. Above the critical Reynolds number quite small perturbations, perhaps introduced at the inlet or by an irregularity of the wall of the tube, grow rapidly with a sinuous motion. Soon they grow so much that nonlinearity becomes strong and large eddies (Figure 1.2(c)) or turbulent spots (Figure 1.3) form. (This mechanism, whereby a flow which is stable to all infinitesimal perturbations is made to change abruptly to a turbulent or nearly turbulent flow by a finite-amplitude perturbation, is now often called *bypass transition*.) As the Reynolds number increases, the threshold amplitude of perturbations to create instability decreases. At high Reynolds numbers turbulence ensues at once due to the inevitable presence of perturbations of small amplitude, and the flow becomes random, strongly three-dimensional (that is, very non-axisymmetric), and strongly nonlinear everywhere.† This instability of Poiseuille pipe flow may be contrasted with that of plane Poiseuille flow, which is unstable to infinitesimal perturbations at sufficiently large values of the Reynolds number. This explanation is supported by the treatment of the theory of the linear stability of Poiseuille pipe flow in §8.10. However, in practice the instability of plane Poiseuille flow resembles the instability of Poiseuille pipe flow, at least superficially (see Figure 1.4).

The physical mechanisms of Reynolds's experiments on instability of Poiseuille flow in a pipe are vividly illustrated by a film loop made by Stewart (FL1968) for the Education Development Center. This loop consists of edited excerpts from his longer film on *Turbulence* (Stewart, F1968). Details of these and other motion pictures on hydrodynamic stability may be found after the list of references at the end of the book. Videos of the experiment can be seen

---

† Many of the features of the transition from laminar to turbulent flow can easily be appreciated by observing the smoke from a cigarette. Light the cigarette, point the burning tip upwards, and observe the smoke as it rises from rest. See also Van Dyke (1982, Fig. 107).

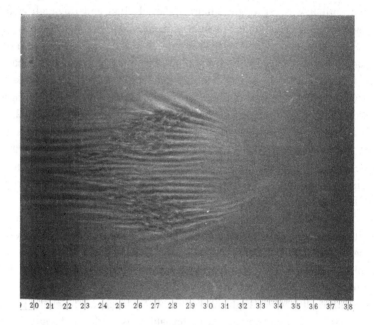

Figure 1.4   A turbulent spot triggered by jets in the wall of plane Poiseuille flow at $R = 1000$, where $R = Vd/\nu$, V is the maximum velocity of the flow, and the walls are separated by a distance $2d$. (From Carlson *et al.*, 1982, Fig. 4.)

by use of the compact disk of Homsy *et al.* (CD2000); this CD is currently more readily available than the film loops or their video versions, although briefer. Under the heading *Video Library* and subheadings 'Reynolds Transition Apparatus' and 'The Reynolds Transition Experiment', some short videos of recent experiments on Reynolds's original apparatus are shown; further experiments can be found under the subheadings 'Pipe Flow', 'Tube Flow' and 'Turbulent Pipe Flow'. Under the heading *Boundary Layers* and subheadings 'Instability, Transition and Turbulence' and 'Instability and Transition in Pipe and Duct Flow' more short videos are available.

1.2 The Methods of Hydrodynamic Stability

It may help at the outset to recognize that hydrodynamic stability has a lot in common with stability in many other fields, such as magnetohydrodynamics, plasma physics, elasticity, rheology, combustion and general relativity. The physics may be very different but the mathematics is similar. The mathematical essence is that the physics is modelled by nonlinear partial differential equations and the stability of known steady and unsteady solutions is examined. Hydrodynamics happens to be a mature subject (the Navier–Stokes equations having been discovered in the first half of the nineteenth century), and a given motion of a fluid is often not difficult to produce and to see in a laboratory, so hydrodynamic stability has much to tell us as a prototype of nonlinear physics in a wider context.

We learn about instability of flows and transition to turbulence by various means which belong to five more-or-less distinct classes:

(1) *Natural phenomena and laboratory experiments.* Hydrodynamic instability would need no theory if it were not observable in natural phenomena, man-made processes, and laboratory experiments. So observations of nature and experiments are the primary means of study. All theoretical investigations need to be related, directly or indirectly, to understanding these observations. Conversely, theoretical concepts are necessary to describe and interpret observations.

(2) *Numerical experiments.* Computational fluid dynamics has become increasingly important in hydrodynamic stability since 1980, as numerical analysis has improved and computers have become faster and gained more memory, so that the Navier–Stokes equations may be integrated accurately for more and more flows. Indeed, computational fluid dynamics has now reached a stage where it can rival laboratory investigation of hydrodynamic stability by simulating controlled experiments.

(3) *Linear and weakly nonlinear theory.* Linearization for small perturbations of a given basic flow is the first method to be used in the theory of hydrodynamic stability, and it was the method used much more than any other until the 1960s. It remains the foundation of the theory. However, weakly nonlinear theory, which builds on the linear theory by treating the leading nonlinear effects of small perturbations, began in the nineteenth century, and has been intensively developed since 1960.

(4) *Qualitative theory of bifurcation and chaos.* The mathematical theory of differential equations shows what flows *may* evolve as the dimensionless parameters, for example the Reynolds number, increase. The succession of bifurcations from one regime of flow to another as a parameter increases cannot be predicted quantitatively without detailed numerical calculations, but the admissible and typical routes to chaos and thence turbulence may be identified by the qualitative mathematical theory. Thus the qualitative theory of dynamical systems, as well as weakly nonlinear analysis, provides a useful conceptual framework to interpret laboratory and numerical experiments.

(5) *Strongly nonlinear theory.* There are various mathematically rigorous methods, notably Serrin's theorem and Liapounov's direct method, which give detailed results for arbitrarily large perturbations of specific flows. These results are usually bounds giving sufficient conditions for stability of a flow or bounds for flow quantities.

The plan of the book is to develop the major concepts and methods of the theory in detail, and then apply them to the instability of selected flows, relating the theoretical to the experimental results. This plan is itemized in the list of contents. First, in this and the next chapter, many concepts and methods will be described, and illustrated by simple examples. Then, case by case, these methods and concepts, together with some others, will be used in the later chapters to understand the stability of several important classes of flows. The theory of hydrodynamic stability has been applied to so many different classes of flow that it is neither possible nor desirable to give a comprehensive treatment of the applications of the theory in a textbook. The choice of applications below is rather arbitrary, and perhaps unduly determined by tradition. However, the choice covers many useful and important classes of flow, and illustrates well the five classes of general method summarized above.

1.3 Further Reading and Looking

It may help to read some of the following books to find fuller accounts of many points of this text. Many of the books are rather out of date, being written before the advent of computers had made much impact on the theory of hydrodynamic stability. (Perhaps computational fluid dynamics has led to the most important advances in recent years, and perhaps the theory of dynamical systems or applications of the theory has led to a wider physical range of new problems.) However, the subject is an old one, with most of the results of enduring importance, so these books are still valuable.

Betchov & Criminale (1967) is a monograph largely confined to the linear theory of the stability of parallel flows, covering numerical aspects especially well. Chandrasekhar (1961) is an authoritative treatise, a treasure house of research results of both theory and experiment. It emphasizes the linear stability of flows other than parallel flows, with influence of exterior fields such as magnetohydrodynamic, buoyancy and Coriolis forces. Its coverage of the literature is unusual, informative and of great interest. Drazin & Reid (1981) is a monograph with a broad coverage of the subject. It has several problems for students, but few of them are easy. Huerre & Rossi (1998) is a set of 'lecture notes', though at an appreciably higher level than this book. It is an account, mostly of linear stability of mostly parallel flows, with good modern coverage of numerical and experimental as well as theoretical results. Joseph (1976) is a monograph which emphasizes nonlinear aspects, especially the energy method, but has a broad coverage of basic flows. Landau & Lifshitz (1987) is a great treatise masquerading as a textbook; it summarizes the physical essentials of hydrodynamic stability with masterly brevity. Lin (1955) is a classic monograph, largely confined to the linear stability of parallel flows of a viscous fluid, the complement of Chandrasekhar's treatise. Schmid & Henningson (2001) is an up-to-date comprehensive research monograph on instability and transition of parallel flows.

We have already referred to pictures to enrich understanding of Reynolds's experiment. Such pictures are, of course, as valuable in the understanding of many other hydrodynamic instabilities. Van Dyke (1982) is a beautiful collection of photographs of flows, including hydrodynamic instabilities. Nakayama (1988) is another fine collection of photographs of flows, including hydrodynamic instabilities. Look at the photographs relevant to hydrodynamic stability, think about them, and relate them to the theory of this book. However, hydrodynamic instability is a dynamic phenomenon, best seen in motion pictures. So, many relevant films, film loops and videos, and the compact disk of Homsy *et al.* (CD2000), are listed in the Motion Picture Index at the end of the list of references. It is appropriate to add some words of caution here. The results of visualization of *un*steady flows are liable to be misinterpreted. Be careful. In particular, make sure that you understand the difference between streamlines, streaklines and particle paths before you jump to too many conclusions.

2

Introduction to the Theory of Steady Flows, Their Bifurcations and Instability

> ... whosoever heareth these sayings ..., and doeth them, ... will liken ... unto a wise man, which built his house upon a rock: And the rain descended, and the floods came, and the wind blew, and beat upon that house; and it fell not: for it was founded on rock.
>
> *Matt. viii 24–25*

The essences of the common forms of bifurcation, that is, the common types of change of regime of flow, are introduced in this chapter by use of simple illustrative ordinary differential problems. It is shown afterwards that these bifurcations occur where instability occurs. Finally, stability of a flow is defined mathematically, and the linearized problem and the method of normal modes are described.

2.1 Bifurcation

Consider flows of an incompressible viscous fluid in a given domain $\mathcal{V}$. Let ρ be the density of the fluid, and ν the kinematic viscosity. Let $\mathbf{u}_*$, p_* be the velocity and pressure of the fluid at a given point $\mathbf{x}_*$ at time t_*. Then flow is governed by the Navier–Stokes equations,

$$\frac{\partial \mathbf{u}_*}{\partial t_*} + \mathbf{u}_* \cdot \nabla_* \mathbf{u}_* = -\frac{1}{\rho}\nabla_* p_* + \nu \Delta_* \mathbf{u}_*,$$

and the equation of continuity,

$$\nabla_* \cdot \mathbf{u}_* = 0,$$

in $\mathcal{V}$; and certain boundary conditions, say

$$\mathbf{u}_* = \mathbf{U}_{0*} \quad \text{on part of } \partial\mathcal{V}, \ \mathbf{u}_* \text{ is periodic on the rest of } \partial\mathcal{V};$$

where Δ_* is the Laplacian operator, $\partial\mathcal{V}$ is the boundary of $\mathcal{V}$ and $\mathbf{U}_{0*}$ is a given velocity of the fluid on the boundary.

Suppose that these equations and boundary conditions have a certain solution, approximate if not exact, which describes a steady flow whose stability is of interest. Let this basic flow have velocity field $\mathbf{U}_*$ and pressure P_*. It will often be convenient to choose dimensionless variables, and define an appropriate Reynolds number R. Let us choose some characteristic length scale L of the basic flow, such as the radius of the domain of flow, and some characteristic velocity scale V, such as the greatest value of $|\mathbf{U}_*|$ in $\mathcal{V}$. For example, for the flow of a uniform stream around a sphere, V might be the velocity of the stream, and L the radius or the diameter of the sphere. Then we may define dimensionless variables such as $\mathbf{x} = \mathbf{x}_*/L$, $t = Vt_*/L$, $\mathbf{u} = \mathbf{u}_*/V$, $p = p_*/\rho V^2$ and so forth, and a Reynolds number $R = VL/\nu$. Now the velocity field $\mathbf{U}(\mathbf{x}, R)$ and pressure field $P(\mathbf{x}, R)$, which specify the basic flow in dimensionless variables, satisfy the Navier–Stokes equations,

$$\mathbf{U} \cdot \nabla \mathbf{U} = -\nabla P + R^{-1} \Delta \mathbf{U}, \qquad (2.1)$$

$$\nabla \cdot \mathbf{U} = 0 \qquad (2.2)$$

in $\mathcal{V}$; and boundary conditions

$$\mathbf{U} = \mathbf{U}_0 \quad \text{on part of } \partial\mathcal{V}; \ \mathbf{U} \text{ is periodic on the rest of } \partial\mathcal{V}. \qquad (2.3)$$

In general $\mathbf{U}$, P depend on R, and there may be more than one steady solution $\mathbf{U}$, P for the same value of R and the same boundary conditions. We shall see that *bifurcation*, that is, change in the number, or in the qualitative character, of the set of possible steady flows (or unsteady flows in dynamic equilibrium) as R varies, is often linked with the onset of instability.

The important physical idea of a succession of instabilities and changes of flow regime, along the 'route to turbulence', as the Reynolds number increases will next be introduced by some very simple model problems of bifurcation. It may seem at first that these models of algebraic and ordinary-differential problems are too simple to be relevant to hydrodynamics. Yet we shall eventually show that not only do they illustrate many important concepts of stability, but they also represent asymptotically the *local* properties of many instabilities and bifurcations of solutions of the Navier–Stokes equations governing the flow of fluids. This is because, although the Navier–Stokes equations are partial differential equations whose solutions belong to an infinite-dimensional function space, a solution may be approximated asymptotically by an element of a low-dimensional space. In practice a solution may be represented by a spectral expansion, such as a Fourier series over space, and approximated asymptotically by only a few components of the expansion, whose amplitudes satisfy an ordinary differential system of low order. This point will be taken up in later chapters, and illustrated with examples of several flows.

Example 2.1: A turning point. Consider, merely as a simple model problem for illustrative purposes, or 'toy' problem, the quadratic equation

$$a - l(U - U_0)^2 = 0,$$

where $a = k(R - R_c)$, for some constants $k > 0, l \neq 0, U_0$ and R_c. A toy problem can be useful in learning about a complicated property of fluid motion if the simple toy problem describes that property, even if it does not describe most other properties of the motion. Here we may regard U as representing a given component of the velocity of the fluid at some given point of a steady flow as a function of the Reynolds number. Then

$$U = U_0 \pm [k(R - R_c)/l]^{1/2}.$$

This gives two solutions when $k(R - R_c)/l > 0$, one when $R = R_c$, and none when $k(R - R_c)/l < 0$. It is convenient to plot solutions in a *bifurcation diagram*, where some variable describing the state of particular flows is plotted against some flow parameter specifying the fluid or the configuration of the flow, and so forth. Here we plot U against R in Figure 2.1 for the case $kl > 0$. We say that there is a *simple turning point, fold* or a *saddle-node bifurcation* at $R = R_c, U = U_0$. It is called a *bifurcation point* because the number (and character) of the solutions changes there. □

Example 2.2: A transcritical bifurcation. As another very simple model of bifurcation of steady solutions of the Navier–Stokes equations, consider next the quadratic equation

$$aU - lU^2 = 0.$$

Therefore

$$U = 0 \quad \text{or} \quad U = a/l, = k(R - R_c)/l,$$

so there are two solutions for all $R \neq R_c$. The bifurcation at $R = R_c, U = 0$ is an example of what is called a *transcritical point*. See Figure 2.2. □

Example 2.3: Pitchfork bifurcation. Next take the model equation

$$aU - lU^3 = 0,$$

which is typical for the first bifurcation of flows with symmetry in $\pm U$. Then

$$U = 0, \quad \text{or} \quad U = \pm [k(R - R_c)/l]^{1/2} \quad \text{if } k(R - R_c)/l > 0.$$

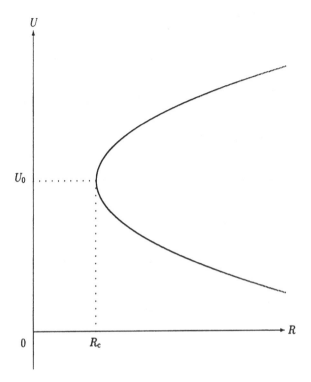

Figure 2.1 The bifurcation diagram for the turning point in the (R, U)-plane for the case $l > 0$.

There is said to be a *pitchfork bifurcation* at $R = R_c$, $U = 0$. We see that there is *symmetry breaking* at $R = R_c$, in the sense that if $kl > 0$, then there is a unique symmetric solution for $R < R_c$, but there is also a pair of asymmetric solutions for $R > R_c$. See Figure 2.3. $\square$

Example 2.4: Plane Couette–Poiseuille flow. If we seek a plane parallel flow of an incompressible viscous fluid, then we assume that the velocity is $\mathbf{U} = U(z)\mathbf{i}$, say, and substitute this velocity into the dimensional vorticity equation for two-dimensional flow in the (x, z)-plane (which is a convenient form of the Navier–Stokes equations for the purpose),

$$\frac{\partial \eta}{\partial t} + u \frac{\partial \eta}{\partial x} + w \frac{\partial \eta}{\partial z} = \nu \Delta \eta,$$

where $\eta = \partial u / \partial z - \partial w / \partial x$. It follows that

$$\frac{\mathrm{d}^3 U}{\mathrm{d} z^3} = 0.$$

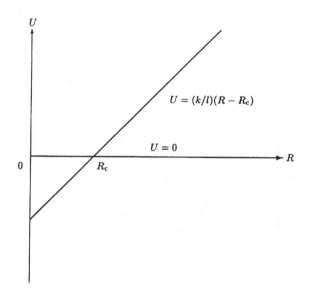

Figure 2.2   The bifurcation diagram for the transcritical point in the (R, U)-plane for the case $l > 0$.

Therefore

$$U(z) = Az^2 + Bz + C$$

for some constants A, B, C to be determined by the pressure gradient and boundary conditions. The pressure gradient along the pipe must be constant for this to be a solution. Also the no-slip condition at each wall must be satisfied. In fact $A = (2\mu)^{-1}\partial p/\partial x$, where μ is the dynamic viscosity of the fluid. In particular we find *plane Couette flow* in a channel with

$$U(z) = Vz/L \quad \text{for } -L \le z \le L$$

if $\partial p/\partial x = 0$ and there are rigid plates at $z = \pm L$ moving with velocities $\pm V$ respectively. In this case the Reynolds number is often chosen as $R = VL/\nu$. Also we find *plane Poiseuille flow* with

$$U(z) = V\left(1 - z^2/L^2\right) \quad \text{for } -L \le z \le L$$

if there are fixed rigid plates at $z = \pm L$, where $V = -(L^2/2\mu)\partial p/\partial x$. Again it is common to define $R = VL/\nu$ for this flow. The configuration of these flows is shown in Figure 2.4.

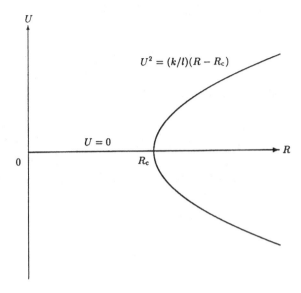

Figure 2.3 The bifurcation diagram for the pitchfork in the (R, U)-plane for the case $l > 0$.

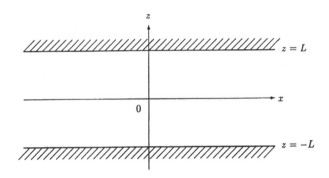

Figure 2.4 The configuration of plane Couette–Poiseuille flows.

Note that the basic parallel flow in this example is both unique and independent of the value of the Reynolds number. This is a special property of parallel flows, and is atypical of flows of a viscous fluid. The property is due to the vanishing of the inertial terms in the Navier–Stokes equation, which in turn is due to the geometrical character of the flow, so that the Reynolds number for dynamically similar flows is *not* the usual characteristic ratio of inertial to viscous forces. Usually a steady basic flow changes as R increases, and is not the unique steady flow in the given circumstances. □

Example 2.5: Jeffery–Hamel flows. For the next example, we choose some steady flows which do change as the Reynolds number increases, and flows which are not unique (see, e.g., Batchelor, 1967, §5.6).

Consider two-dimensional flow of an incompressible viscous fluid between two rigid planes with equations $\theta = \pm\alpha$ driven by a steady line source (or sink) of volume flux Q, per unit distance normal to the plane of flow, at the intersection $r_* = 0$ of the two planes, where (r_*, θ) are plane polar coordinates. The configuration and the coordinates are indicated in Figure 2.5. Therefore the boundary conditions are that

$$\psi_* = \pm\frac{1}{2}Q, \qquad \partial\psi_*/\partial\theta = 0 \quad \text{at } \theta = \pm\alpha,$$

where ψ_* is the streamfunction, such that the radial velocity component $u_{r*} = \partial\psi_*/r_*\partial\theta$ and the transverse component $u_\theta = -\partial\psi_*/\partial r_*$.

Now seek steady flows for which the streamfunction depends only on θ, that is, for which $\psi_* = \Psi_*(\theta)$. This gives purely radial flow with velocity $u_{r*} = U_* = d\Psi_*/r_*d\theta$, and vorticity $\zeta_* = -\Delta_*\psi_* = -d^2\Psi_*/r_*^2d\theta^2$.

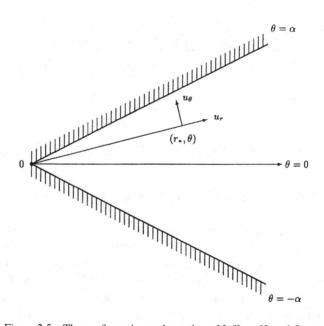

Figure 2.5   The configuration and notation of Jeffery–Hamel flows.

Next choose dimensionless variables with $\psi = \psi_*/\frac{1}{2}Q$ and so forth, and define the Reynolds number as $R = Q/2\nu$. Then the dimensionless form of the vorticity equation becomes

$$\frac{\partial \zeta}{\partial t} + \frac{1}{r}\frac{\partial(\zeta, \psi)}{\partial(r, \theta)} = R^{-1}\Delta\zeta, \tag{2.4}$$

which, with $\psi = \Psi(\theta)$, can be reduced to give the nonlinear *ordinary* differential equation

$$\frac{d^4\Psi}{d\theta^4} + 4\frac{d^2\Psi}{d\theta^2} + 2R\frac{d\Psi}{d\theta}\frac{d^2\Psi}{d\theta^2} = 0. \tag{2.5}$$

The boundary conditions give

$$\Psi = \pm 1, \qquad d\Psi/d\theta = 0 \quad \text{at } \theta = \pm\alpha. \tag{2.6}$$

This nonlinear boundary-value problem can be solved in explicit terms of Jacobian elliptic functions, but it is in most respects more easily solved numerically. There is a rich variety of solutions, of types I, II_n, III_n, IV_n and V_n, in the nomenclature of Fraenkel (1962, p. 124); some velocity profiles are sketched in Figure 2.6. (You need not learn the precise meanings of the types

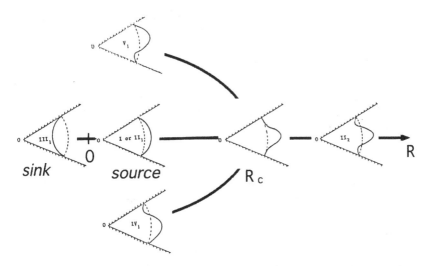

Figure 2.6   Sketches of the velocity profiles of the more important Jeffery–Hamel flows. (After P. Drazin & T. Kambe, *Ryutai Rikigaku – Anteisei To Ranyu* (*Fluid Dynamics – Stability and Turbulence*), University of Tokyo Press, 1989, Fig. 2.4. Reproduced by permission of the University of Tokyo Press.)

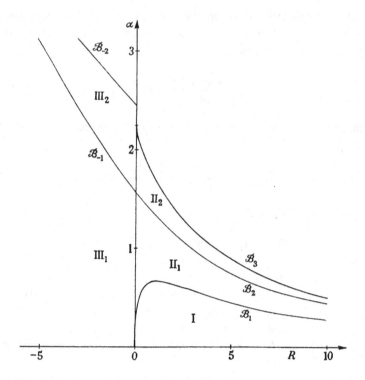

Figure 2.7   The regions of occurrence of the most important types of Jeffery–Hamel flow in the (R, α)-plane. Here $R > 0$ corresponds to a line source (with diverging net flow for $Q > 0$) and $R < 0$ corresponds to a line sink (with converging flow) at $r = 0$. (After Fraenkel, 1962, Fig. 5; reproduced by permission of the Royal Society.)

and subscripts, because they are unimportant for the present purpose of an illustrative example.)

At the risk of over-simplification, the Jeffery–Hamel flows may be summarized as follows. For any given pair of values of (R, α) there is an infinity of possible steady solutions. The most important (including all the stable) solutions are shown below both in the (R, α)-plane (Figure 2.7) and in the bifurcation diagram (Figure 2.8) in the $(\alpha, \Psi'(0))$-plane for a given 'typical' value of R. The main thing to bear in mind about these diagrams is not the quantitative details but the fact that there is a complicated structure with many turning points and pitchforks. Note also that inflow is more stable than outflow, other things being equal. The evidence to make plausible the assertions of which solutions are stable and which unstable is sketched in §10.3.3.

(We could see that $\mathcal{B}_2$ is indeed a pitchfork bifurcation with II_1 as the 'handle', II_2 as the middle 'prong', and IV_1, V_1 as the side 'prongs', better

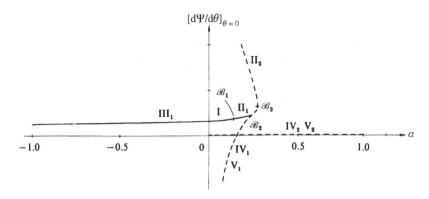

Figure 2.8 The bifurcation diagram of Jeffery–Hamel flows in the $(\alpha, [d\Psi/d\theta]_{\theta=0})$-plane for $R = 20$ (after Sobey & Drazin, 1986, Fig. 3). It may help to regard this diagram as the projection on to the plane of a curve in a three-dimensional space, each branch of the curve representing a Jeffery–Hamel solution. Here $\alpha > 0$ corresponds to a line source and, by convention, $\alpha < 0$ to a sink. Note that two solutions of type II$_2$ may occur between the pitchfork bifurcation $\mathcal{B}_2$ (visualize it 'sideways') and the turning point $\mathcal{B}_3$. A stable solution is denoted by a continuous curve and an unstable solution by a broken curve.

in the $(\alpha, [d^2\Psi/d\theta^2]_{\theta=0})$-plane, because IV$_1$ and V$_1$ have the same velocity u_r at the centre line $\theta = 0$; however, the symmetric flows I, II$_n$, III$_n$ all have $d^2\Psi/d\theta^2 = 0$ at $\theta = 0$.) $\square$

2.2 Instability

The complexity of flows of a viscous fluid, and so of the solutions of the Navier–Stokes equations, leads us to present model problems of simple ordinary-differential equations in this section in order to explain the fundamental ideas of stability. We have already introduced some steady solutions and their bifurcations; next we shall consider the instabilities of these solutions, and find to what other solutions, steady or unsteady, they may evolve when subject to small perturbations. By way of introduction, the previous section describes only a few of the simplest bifurcations. More general ordinary-differential systems both have bifurcations which resemble these simple ones locally (as a general smooth curve with a vertical tangent at a point resembles a parabola with the same tangent and same curvature at the point), and have more complicated forms of bifurcation. Further, more general systems often have sequences of bifurcations as a parameter, such as the Reynolds number, increases; this important evolution of solutions as a parameter increases can be seen in Example 2.5 and in Exercise 6.11 and will be elaborated in Chapter 9.

Real hydrodynamic problems usually need a lot of numerical calculation and so are less instructive than simple model problems; some hydrodynamic problems with a strong symmetry (for example, plane parallel flows are symmetric with respect to the group of translations in the direction of flow) may be solved by reducing the stability problem to an ordinary-differential one, and a few of these stability problems have explicit solutions – of course, these simple (and atypical) solutions are those which appear most often in textbooks and lecture courses.

Example 2.6: A turning point again. Take the model equation,

$$\frac{du}{dt} = a - l(u - U_0)^2, \tag{2.7}$$

again with $a = k(R - R_c), k > 0$. By using the easily obtained explicit analytic solution of this ordinary differential equation, or by considering qualitatively the sign of du/dt (and hence whether u increases or decreases as t increases), it can be shown that any small perturbation of the steady solution $u = U_+, = U_0 + (a/l)^{1/2}$, will decay as $t \to \infty$ and hence that the solution is stable. Similarly, some small perturbations of $u = U_-, = U_0 - (a/l)^{1/2}$, will grow so that eventually they are no longer small, and hence that the solution is said to be unstable. Taking $l > 0$, we see that if $R > R_c$ then a small initial perturbation of U_- gives $u(t) \to U_+$ as $t \to \infty$ or $u(t) \to -\infty$ as $t \to$ a finite limit (in fact) according to the sign of the perturbation, and a small initial perturbation of U_+ gives $u(t) \to U_+$ as $t \to \infty$ (see Figure 2.9). $\square$

Example 2.7: A transcritical bifurcation again. Next take

$$\frac{du}{dt} = au - lu^2, \tag{2.8}$$

where $a = k(R - R_c), k > 0$. Again we may find the explicit analytic solution or use qualitative methods to show that the solutions are as sketched in Figure 2.10 in the case when $l > 0$. $\square$

Example 2.8: A pitchfork bifurcation again. Next take

$$\frac{du}{dt} = au - lu^3, \tag{2.9}$$

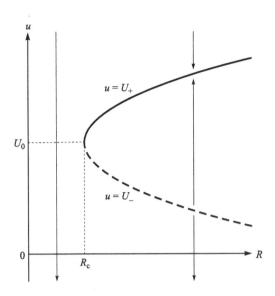

Figure 2.9 The bifurcation diagram in the (R, u)-plane for the example of a turning point for the case $l > 0$. The stable steady solution U_+ is denoted by a continuous curve, and the unstable steady solution U_- by the broken curve. The vertical lines with arrows indicated how a time-dependent solution u varies as t increases for fixed R.

where $a = k(R - R_c), k > 0$. This is a *Landau equation*, essentially the one first proposed as a model of hydrodynamic stability by Landau in 1944. The equation is unchanged in form if we change the sign of u, so it often appears as the weakly nonlinear equation governing the amplitude of the most unstable (or least stable) eigenfunction for stability of a flow with such a symmetry (this might be mirror symmetry of the configuration of flow about some plane, or the translational symmetry where the amplitude of a wave at a point is the negative of the amplitude half a wavelength away). It is the classic prototype of symmetry breaking. We shall treat the equation by the same methods as before, distinguishing two cases according to the sign of the *Landau constant* l.

For the case $l > 0$ we see *supercritical stability*, that is, the two stable solutions for R greater than its critical value R_c for linear stability in addition to the unstable *solution* $u = 0$. Note that $u(t) \rightarrow [\text{sgn}u(0)][k(R - R_c)/l]^{1/2}$ as $t \rightarrow \infty$ if $R > R_c$ whereas $u(t) \rightarrow 0$ as $t \rightarrow \infty$ if $R \leq R_c$; in the former case the ultimate state depends only on the sign of the initial value $u(0)$ of U, not its magnitude, and in the latter case the ultimate state is the same for all initial values. Some typical solutions are sketched in Figure 2.11(a).

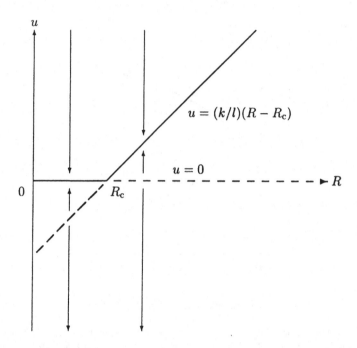

Figure 2.10   Sketch of the bifurcation diagram in the (R, u)-plane for the example of a transcritical bifurcation for the case $l > 0$.

For the case $l < 0$, we see two unstable solutions for $R < R_c$ in addition to the stable solution $u = 0$. Note that there is a 'threshold' such that if $|u(0)| < [k(R - R_c)/l]^{1/2}$, then $u(t) \to 0$ as $t \to \infty$, but if $|u(0)| > [k(R - R_c)/l]^{1/2}$, then $u(t)$ increases to infinity monotonically as t increases. You can show, by solving the differential equation explicitly, that $u(t) \to [\text{sgn}u(0)]\infty$ as $t \to$ a *finite* number (according to this model, at any rate) which depends on $u(0)$ as well as k, R, l. Some typical solutions are sketched in Figure 2.11(b).

We can confirm some of these results by looking at small perturbations of the steady solutions. Define the *perturbation* of U for a given solution u as

$$u'(t) = u(t) - U,$$

where U is one of the steady solutions of $aU = lU^3$. (Perturbations are often called *disturbances*.) Then

$$\frac{\mathrm{d}u'}{\mathrm{d}t} = \frac{\mathrm{d}u}{\mathrm{d}t} = au - lu^3$$
$$= a(U + u') + l(U + u')^3$$

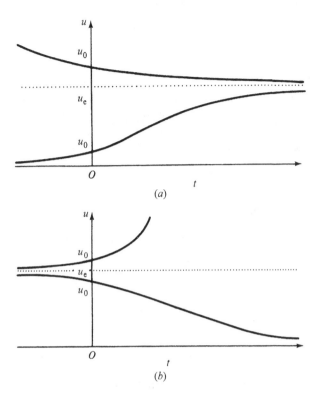

(a)

(b)

Figure 2.11 (a) Supercritical stability for $0 < R - R_c \ll 1$ and $l > 0$: the development of the solution u of the Landau equation (2.9) as a function of time for a few initial values u_0. (b) Subcritical stability for $0 < R - R_c \ll 1$ and $l > 0$: the development of the solution u of the Landau equation (2.9) as a function of time for a few initial values u_0. In each diagram $u_e = [k(R - R_c)/l]^{1/2}$. (After Drazin & Reid, 1981, Figs. 7.1(a), 7.2(a).)

$$= aU - lU^3 + (a - 3lU^2)u' + O(u'^2) \quad \text{as } u' \to 0$$
$$= (a - 3lU^2)u' + O(u'^2),$$

because U is one of the steady solutions. In studying stability, we study the growth of solutions near the given solution U, so we may plausibly linearize, and consider

$$\frac{du'}{dt} = (a - 3lU^2)u'$$

For the null solution $U = 0$, this gives

$$\frac{du'}{dt} = au'.$$

Therefore

$$u'(t) = u(0)e^{st},$$

where the exponent $s = a, = k(R - R_c)$ for $k > 0$ (in realistic applications). Then there is linear stability, with exponential decay, if $R < R_c$ and linear instability, with exponential growth, if $R > R_c$. This solution of the linearized equation which grows exponentially with time is an example of a *normal mode*.

If $U = \pm[k(R - R_c)/l]^{1/2}$ for $l > 0, R > R_c$, then we similarly find $u'(t) = u(0)e^{st}$, but where now

$$s = a - 3lU^2 = -2k(R - R_c) < 0$$

and so gives supercritical stability, as indicated in Figure 2.12(a).

All these results may be confirmed by use of the 'exact' explicit solution of the Landau equation. $\square$

Example 2.9: A Hopf bifurcation. Consider

$$\frac{dx}{dt} = -y + (a - x^2 - y^2)x, \qquad \frac{dy}{dt} = x + (a - x^2 - y^2)y,$$

where $a = k(R - R_c), k > 0$. The only steady solution of this system is the null solution $x = y = 0$. To find its stability we linearize the system with respect to small perturbations of the null solution, finding

$$\frac{dx}{dt} = -y + ax, \qquad \frac{dy}{dt} = x + ay.$$

We solve this linearized system by again using the *method of normal modes*, that is, by supposing that $x, y \propto e^{st}$, and deducing that

$$sx = ax - y, \qquad sy = x + ay,$$

and therefore that s is an eigenvalue of the matrix

$$\mathbf{J} = \begin{bmatrix} a & -1 \\ 1 & a \end{bmatrix}.$$

Therefore

$$0 = \det(\mathbf{J} - s\mathbf{I}) = (a - s)^2 + 1.$$

Therefore

$$s = a \pm i = k(R - R_c) \pm i.$$

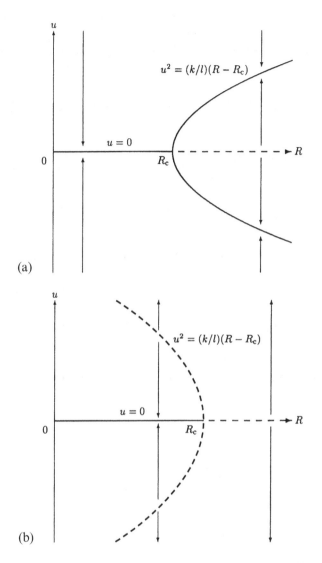

Figure 2.12   Bifurcation diagrams in the (R, u)-plane for the Landau equation: (a) supercritical stability, $l > 0$; (b) subcritical instability, $l < 0$.

Therefore

$$x(t) = \frac{1}{2}\big(Ae^{it} + A^*e^{-it}\big)e^{at}, \qquad y(t) = -\frac{1}{2}i\big(Ae^{it} - A^*e^{-it}\big)e^{at}$$

for some complex constant A, which may be determined by use of the initial conditions, where an asterisk is used as a superscript to denote complex conjugation. This gives stability, with exponential decay, if $\mathrm{Re}(s) < 0$ for both eigenvalues, that is, if $R < R_c$, and similarly instability if $R > R_c$.

In fact it is informative to transform to polar coordinates r, θ, where $r \geq 0$, $x = r\cos\theta$, $y = r\sin\theta$, in which the system decouples as

$$\frac{\mathrm{d}r}{\mathrm{d}t} = r(a - r^2), \qquad \frac{\mathrm{d}\theta}{\mathrm{d}t} = 1,$$

and thence to find the exact solution. The solution (in Example 2.8) implies that $r(t) \to 0$ as $t \to \infty$ for all $r(0)$ if $R \leq R_c$ and $r(t) \to a^{1/2} = [k(R - R_c)]^{1/2}$ as $t \to \infty$ for all $r(0)$ if $R > R_c$. Also $\theta(t) = \theta_0 + t$ for all $\theta(0) = \theta_0$. This gives, for all $R > R_c$, a nonlinear solution $x = r\cos\theta$, $y = r\sin\theta$ of period 2π as $t \to \infty$. Such a periodic solution of a differential equation which is approached by neighbouring solutions as time increases is called a *limit cycle*. Two typical orbits in the phase plane of (x, y), as t increases, are shown in Figure 2.13 for the case $R > R_c$; note how the limit cycle attracts neighbouring orbits.

This example is typical of *Hopf bifurcations*, in which the real part $\mathrm{Re}(s)$ of a complex conjugate pair of eigenvalues increases through zero as a parameter increases or decreases through a critical value, here as R increases through R_c, and an oscillatory solution bifurcates from the steady solution where it becomes unstable. Of course, it is no accident that a real system often has a complex conjugate pair of eigenvalues, so we meet Hopf bifurcations for partial

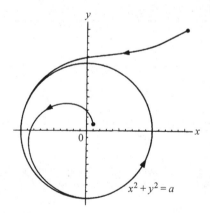

Figure 2.13 Two orbits in the (x, y)-plane for the system $\mathrm{d}x/\mathrm{d}t = \mathrm{d}x/\mathrm{d}t = -y + (a - x^2 - y^2)x$, $\mathrm{d}y/\mathrm{d}t = x + (a - x^2 - y^2)y$ of Example 2.9 when $R > R_c$. (After P. Drazin & T. Kambe, *Ryutai Rikigaku – Anteisei To Ranyu (Fluid Dynamics – Stability and Turbulence)*, University of Tokyo Press, 1989, Fig. 2.10. Reproduced by permission of the University of Tokyo Press.)

differential systems governing flows as well as for this simple example of an ordinary differential equation. So it is important to determine from the linear problem whether the exponent s is zero or purely imaginary at the margin of stability: in the former case a turning point, a transcritical or pitchfork bifurcation typically occurs, and in the latter case a Hopf bifurcation. In the *fluid dynamical context*, it is sometimes said that the *principle of exchange of stabilities* is valid when the time exponent of the least stable normal mode is zero at the margin of stability. $\square$

Before moving on, note that we have used a complex representation of a real solution of a real problem in Example 2.9. This idea, based on the property that if a complex function satisfies a real homogeneous equation, then the real and imaginary parts of the function satisfy the equation separately, will be exploited often in the pages that follow. We shall write the complex solution of a real linearized equation or system of equations, meaning implicitly that its real part represents the appropriate physical quantity such as a perturbation of a velocity component or the pressure; for example, we may write $x(t) = Ae^{(a+\mathrm{i})t}$, where A is some complex constant, to mean its real part $\frac{1}{2}(Ae^{\mathrm{i}t} + A^*e^{-\mathrm{i}t})e^{at} = |A|e^{at}\cos(t + \arg A)$. This is the traditional way to use the method of normal modes.

These examples have been chosen for their simplicity rather than to illustrate all aspects of hydrodynamic stability. One common phenomenon they do not illustrate is the instability of the supercritically stable bifurcated flow itself as the Reynolds number increases substantially above the critical value for a pitchfork or Hopf bifurcation. Then we call the first flow the *primary flow*, its instability the *primary instability*, the supercritically stable flow the *secondary flow* and its instability the *secondary instability*. These successive instabilities are discussed further in §9.1.

This section as a whole serves to introduce some important concepts (basic solution, stability, bifurcation) and methods (linearization, normal modes) of the theory of hydrodynamic stability by use of simple ordinary differential equations. Ordinary differential equations will be used later to illustrate other important concepts (such as quasi-periodic solutions and chaos) and methods (weakly nonlinear perturbation) of hydrodynamic stability. However, it should not be forgotten that the motion of a fluid involves space as well as time, and that it is modelled by *partial* differential equations. This means that the use of ordinary differential models is limited, albeit valuable pedagogically. More realistic models with the partial differential equations of hydrodynamics are treated in the next section, which covers some fundamental concepts and methods of the theory of hydrodynamic stability, especially the linear theory.

2.3 Stability and the Linearized Problem

First select a *basic flow* of interest, that is, a solution of the governing equations of motion, whose stability we wish to investigate. This solution may be easy or hard to find; it may be known explicitly in analytic terms, or known only numerically. For example, we may specify the basic flow by the velocity field $\mathbf{U}(x, t)$ and pressure field $P(\mathbf{x}, t)$ of an incompressible viscous fluid in a given domain $\mathcal{V}$ with boundary $\partial\mathcal{V}$. This flow is governed by the Navier–Stokes equations. Then, in dimensionless variables,

$$\frac{\partial \mathbf{U}}{\partial t} + \mathbf{U} \cdot \nabla \mathbf{U} = -\nabla P + R^{-1}\Delta\mathbf{U} \tag{2.10}$$

and

$$\nabla \cdot \mathbf{U} = 0 \tag{2.11}$$

in $\mathcal{V}$; and $\mathbf{U} = \mathbf{U}_0$ on one part, and is periodic on the rest, of $\partial\mathcal{V}$; where R is a Reynolds number.

Now, for general initial values $\mathbf{u}(\mathbf{x}, 0)$ of the velocity and $p(\mathbf{x}, 0)$ of the pressure there is a *total flow* with velocity $\mathbf{u}(\mathbf{x}, t)$ and pressure $p(\mathbf{x}, t)$ for $t > 0$ such that

$$\frac{\partial \mathbf{u}}{\partial t} + \mathbf{u} \cdot \nabla \mathbf{u} = -\nabla p + R^{-1}\Delta\mathbf{u}, \tag{2.12}$$

$$\nabla \cdot \mathbf{u} = 0 \tag{2.13}$$

in $\mathcal{V}$; and

$$\mathbf{u} = \mathbf{U}_0 \quad \text{and so forth} \quad \text{on } \partial\mathcal{V}. \tag{2.14}$$

It is convenient to define the *perturbation quantities* $\mathbf{u}' = \mathbf{u} - \mathbf{U}$ and $p' = p - P$, whether they are small or not. Therefore, subtracting corresponding equations above, we deduce, without approximation, that

$$\frac{\partial \mathbf{u}'}{\partial t} + \mathbf{u}' \cdot \nabla \mathbf{U} + \mathbf{U} \cdot \nabla \mathbf{u}' + \mathbf{u}' \cdot \nabla \mathbf{u}' = -\nabla p' + R^{-1}\Delta\mathbf{u}', \tag{2.15}$$

$$\nabla \cdot \mathbf{u}' = 0 \tag{2.16}$$

in $\mathcal{V}$; and

$$\mathbf{u}' = \mathbf{0} \quad \text{on part, and } \mathbf{u}' \text{ is periodic on the rest, of } \partial\mathcal{V}. \tag{2.17}$$

Example 2.10: Poiseuille pipe flow. An exact solution of the Navier–Stokes equations above is, in *dimensional* form,

$$\mathbf{U}_* = V\left(1 - r_*^2/a^2\right)\mathbf{i}, \qquad P_* = p_{0*} - 4\rho\nu V x_*/a^2,$$

where the domain of flow is $V = \{\mathbf{x}_*: 0 \le r_* < a,\ 0 \le \theta < 2\pi,\ -\infty < x_* < \infty\}$ and cylindrical polar coordinates (x_*, r_*, θ) are used. This is the steady flow along a pipe of radius a driven by a pressure gradient $4\rho\nu V/a^2$, the flow studied by Reynolds (1883). We may choose dimensionless variables $r = r_*/a,\ x = x_*/a,\ \mathbf{U} = \mathbf{U}_*/V,\ p = p_*/\rho V^2$ to get

$$\mathbf{U} = \left(1 - r^2\right)\mathbf{i}, \qquad P = p_0 - 4x/R,$$

where $R = Va/\nu$. For boundary conditions we take

$$\mathbf{u} = \mathbf{0} \quad \text{on } r = 1, \qquad \mathbf{u} \to \mathbf{U} \quad \text{as } x \to \pm\infty,$$

with $V = \{\mathbf{x}: 0 \le r < 1\}$ in dimensionless form. Alternatively, we might model a given flow by taking

$$\mathbf{u} = \mathbf{0} \quad \text{at } r = 1,\ \mathbf{u} \text{ has period } L \text{ in } x,$$

with $V = \{\mathbf{x}: 0 \le r < 1,\ 0 \le \theta < 2\pi,\ 0 \le x < L\}$ and $\partial V = \{\mathbf{x}: r = 1 \text{ for } 0 \le x \le L \text{ or } x = 0, L \text{ for } 0 \le r \le 1\}$ for given $L > 0$. We assume, of course, that $\mathbf{u}$ has period 2π in θ. □

We say, in plain words, that a given basic flow is stable if all perturbations which are small initially remain small for all time, and it is unstable if at least one perturbation which is small initially grows so much that it ceases to remain small after some time. To define 'stability' mathematically we need to specify some metric to give meaning to 'small'. This has been done in many similar ways. We formalize the definition as follows, in the sense of Liapounov.

A basic flow is *stable* if, for all $\epsilon > 0$ there exists $\delta(\epsilon)$ such that if

$$\|\mathbf{u}'(\mathbf{x}, 0)\|,\ \|p'(\mathbf{x}, 0)\| < \delta \tag{2.18}$$

then

$$\|\mathbf{u}'(\mathbf{x}, t)\|,\ \|p'(\mathbf{x}, t)\| < \epsilon \quad \text{for all } t > 0. \tag{2.19}$$

Here the norm $\| \cdots \|$ might be chosen in different ways (and thereby give slightly different definitions of stability); for example, we could choose $\|\mathbf{u}'(\mathbf{x}, t)\| = \sup_{\mathbf{x} \in V} |\mathbf{u}'(\mathbf{x}, t)|$ or $[\int_V \mathbf{u}'^2\, d\mathbf{x}]^{1/2}$ at each instant. Stability with $\|\mathbf{u}'\| \propto [\int_V \mathbf{u}'^2\, d\mathbf{x}]^{1/2}$ is sometimes called *stability in the mean*.

The flow is said to be *asymptotically stable* if it is stable and, moreover,

$$\|\mathbf{u}'(\mathbf{x}, t)\|,\ \|p'(\mathbf{x}, t)\| \to 0 \quad \text{as } t \to \infty. \tag{2.20}$$

In the theory of dynamical systems, an asymptotically stable solution, whether it be steady or unsteady, is called an *attractor*; so in Example 2.9 the null solution is the attractor for $R < R_c$ and the limit cycle is the attractor for $R > R_c$. A system may have more than one attractor, as in Example 2.8 for $R > R_c$ and $l > 0$.

The definition of stability crucially concerns the evolution of *small* perturbations with time, so it is plausible that stability may be investigated by neglecting products of the perturbed quantities in the equations of motion and boundary conditions. This gives the *linearized problem*. From the equations above we find

$$\mathbf{u}'_t + \mathbf{U} \cdot \nabla \mathbf{u}' + \mathbf{u}' \cdot \nabla \mathbf{U} = -\nabla p' + R^{-1} \Delta \mathbf{u}', \tag{2.21}$$

$$\nabla \cdot \mathbf{u}' = 0 \tag{2.22}$$

in $\mathcal{V}$; and

$$\mathbf{u}' = \mathbf{0} \quad \text{or is periodic on } \partial\mathcal{V}. \tag{2.23}$$

If the basic flow is steady, that is, if $\mathbf{U}$ is independent of t, then the linearized problem has coefficients independent of t. It follows plausibly that we may separate the variables, so that the general solution of an initial-value problem is a linear superposition of *normal modes*, each of the form

$$\mathbf{u}'(\mathbf{x}, t) = e^{st}\hat{\mathbf{u}}(\mathbf{x}), \qquad p'(\mathbf{x}, t) = e^{st}\hat{p}(\mathbf{x}), \tag{2.24}$$

where the eigenvalue s and corresponding eigenfunctions $\hat{\mathbf{u}}$, $\hat{p}$ can be found in principle by solving the resultant equations and boundary conditions, namely,

$$s\hat{\mathbf{u}} + \mathbf{U} \cdot \nabla\hat{\mathbf{u}} + \hat{\mathbf{u}} \cdot \nabla\mathbf{U} = -\nabla\hat{p} + R^{-1}\Delta\hat{\mathbf{u}}, \tag{2.25}$$

$$\nabla \cdot \hat{\mathbf{u}} = 0 \tag{2.26}$$

in $\mathcal{V}$, and

$$\hat{\mathbf{u}} = \mathbf{0} \quad \text{or is periodic on } \partial\mathcal{V}. \tag{2.27}$$

(The separation of the variable t can be motivated, and in many cases justified, by use of a Laplace transform.) The eigenvalues s of this real problem are real or occur in complex conjugate pairs. If $\mathcal{V}$ is bounded, then there is a countable infinity of discrete eigenvalues. We deduce that the basic flow is stable if $\text{Re}(s) < 0$ for *all* the eigenvalues s and unstable if $\text{Re}(s) > 0$ for at least one eigenvalue s, because the mode grows in time like

$$\exp[\text{Re}(s)t + i\,\text{Im}(s)t] = \exp[\text{Re}(s)t] \times \{\cos[\text{Im}(s)t] + i\sin[\text{Im}(s)t]\}.$$

(In the case $\text{Re}(s) = 0$ of *neutral stability* according to the linear theory, nonlinear terms *may* render the flow unstable.) For one normal mode the spatial structure of the perturbation does not change as the mode travels, grows or decays with time; although, of course, the structure of a superposition of more than one mode does.

It should be noted that for some basic flows the general solution of the initial-value problem cannot be expressed as a superposition of exponentially growing or decaying normal modes. Exercise 2.8(i) illustrates this point.

Note also that, if the eigenvalue s is complex, then the form (2.24) of the solution of a real problem must be interpreted as implicitly meaning its real part. This is permissible because complex eigenvalues of a real system occur in complex conjugate pairs, and the real and imaginary parts of the solution of a real linearized system satisfy the system separately.

We define the *critical Reynolds number* R_c, say, such that if $R \leq R_c$, then $\mathrm{Re}(s) \leq 0$ for all eigenvalues and that $\mathrm{Re}(s) > 0$ for at least one mode and for at least one value of R in any neighbourhood of R_c. Then we say that the basic flow is *marginally stable* when $R = R_c$, because the flow is stable if $R < R_c$ and unstable for all small enough positive values of $R - R_c$; in practice, flows usually become more unstable as they get faster, so we usually expect to find instability for all $R > R_c$ in this event. Marginal stability implies neutral stability. Neutral and marginal stability almost always coincide when the fluid is viscous, but for inviscid fluids they usually do not, and so it is often useful to distinguish between the two.

*If there is a countable complete set of normal modes with, say, eigenfunctions $\hat{\mathbf{u}}_\mathbf{n}$, $\hat{p}_n$ belonging to eigenvalue s_n for $n = 0, 1, \ldots$, then any initial perturbation can be expressed as a superposition of the modes, say

$$\mathbf{u}'(\mathbf{x}, 0) = \sum_{n=1}^{\infty} a_n \hat{\mathbf{u}}_\mathbf{n}(\mathbf{x}), \qquad p'(\mathbf{x}, 0) = \sum_{n=1}^{\infty} b_n \hat{p}_n(\mathbf{x}), \tag{2.28}$$

for some coefficients a_n, b_n. Now the eigenvalue problem is real, so that each eigenfunction is either real or one of a complex conjugate pair. Therefore each coefficient, a_n or b_n, is either real or one of a complex conjugate pair. The modes may be ordered so that $\mathrm{Re}(s_1) \geq \mathrm{Re}(s_2) \geq \cdots$. It follows that

$$\mathbf{u}'(\mathbf{x}, t) = \sum_{n=1}^{\infty} a_n \hat{\mathbf{u}}_n(\mathbf{x}) \exp(s_n t), \qquad p'(\mathbf{x}, t) = \sum_{n=1}^{\infty} b_n \hat{p}_n(\mathbf{x}) \exp(s_n t), \tag{2.29}$$

Therefore

$$\mathbf{u}'(\mathbf{x}, t) \sim a_1 \hat{\mathbf{u}}_1(\mathbf{x}) \exp(s_1 t), \qquad p'(\mathbf{x}, 0) \sim b_1 \hat{p}_1(\mathbf{x}) \exp(s_1 t) \quad \text{as } t \to \infty \tag{2.30}$$

if there is a unique fastest growing or slowest decaying mode, that is, $s_1 >$ $\mathrm{Re}(s_2) \geq \mathrm{Re}(s_3) \geq \cdots$; or

$$\mathbf{u}'(\mathbf{x}, t) \sim a_1 \hat{\mathbf{u}}_1(\mathbf{x}) \exp(s_1 t) + a_1^* \hat{\mathbf{u}}_1^*(\mathbf{x}) \exp(s_1^* t),$$

$$p'(\mathbf{x}, 0) \sim b_1 \hat{p}_1(\mathbf{x}) \exp(s_1 t) + b_1^* \hat{p}_1^*(\mathbf{x}) \exp(s_1^* t) \quad \text{as } t \to \infty, \qquad (2.31)$$

if there is a conjugate complex pair of fastest growing or slowest decaying modes, that is $\mathrm{Re}(s_1) = \mathrm{Re}(s_2) > \mathrm{Re}(s_3) \geq \cdots$. If the eigenfunctions are complete but not countable, then the superposition may involve an integral, for example a Fourier integral, rather than a sum and there may be modes growing only infinitesimally slower than the fastest growing mode.

*The classic method of normal modes which we have just described and the examples of it that follow in the next chapters may overemphasize the importance of the fastest growing unstable mode. Exponentially growing modes are believed to be so ephemeral that they are rarely observed in natural phenomena or laboratory experiments. One reason for the lack of observations of exponential growth of perturbations is that an unstable basic flow is not set up instantaneously in practice, but rather a stable basic flow evolves slowly until it becomes unstable and then is broken up rapidly by naturally occurring perturbations. Another reason is that in general a perturbation is not the fastest growing mode but a superposition of all modes, stable as well as unstable, each decaying or growing exponentially at a different rate, so that the perturbation does not grow exponentially with time. Indeed, a superposition of exponentially decaying modes may grow by several orders of magnitude *for a while*. (Exercise 2.8(ii) illustrates this point.) According to the *linearized* theory the perturbation grows exponentially (2.30) only after a long time, when the fastest growing mode dominates the others, but in practice perturbations are not very small and nonlinearity usually becomes significant before this stage of exponential growth is reached. Almost all perturbations of an unstable steady or time-periodic basic flow observed in practice are nonlinear. However, the *criterion* for stability of a basic flow given by the linear theory of normal modes is valid in practice (except when subcritical instability occurs). Also exponential growth of controlled perturbations can be seen in numerical simulations of flows.

*We appear to ignore the definition (2.18), (2.19) of stability in practice, because we ascertain the stability of a basic flow merely by finding the eigenvalues of the linearized problem, and deeming the flow stable if no eigenvalue has a positive real part. This is justified by resolving two mathematical issues, which may provoke the thoughts of the more theoretically inclined reader and may safely be ignored by the more practically inclined reader, who is therefore advised to pass on at once to the next paragraph but one. The first issue is whether the basic flow is stable if and only if the null solution of the linearized system is stable, that is, whether the stability of the nonlinear problem is governed by the stability of the linearized problem. A few simple counter-examples have been given to indicate that the basic flow may be unstable when the linearized system is neutrally stable. The second issue is whether the stability of the null solution of the linearized system is governed by the spectrum of the normal modes. It is far from clear that $\sigma_m = \omega_m$ always, where we define

$$\sigma_m = \sup\{\mathrm{Re}(s)\colon s \text{ is an eigenvalue}\}, \quad \omega_m = \sup\left(\lim_{t\to\infty}\{[\log\|\mathbf{u}'(\mathbf{x},t)\|]/t\}\right),$$

on taking the latter supremum over the solutions of the linearized problem for all smooth initial conditions. Much has been written about both these issues, and many counter-examples found to the simple belief that the eigenvalues determine the stability of both the linearized and original problems for nonlinear partial differential systems. However, this simple belief seems adequate to solve problems of stability of flows of a Newtonian fluid, it being accepted that if the null solution of the linearized system is asymptotically stable, then the basic flow is stable, although the basic flow may be unstable when the linearized system is neutrally stable, and that the linearized system is stable if $\sigma_m < 0$. So we shall mostly regard stability of a flow as being determined by the eigenvalues of the linearized problem.

*These issues are not merely pathological conundrums in the mathematical theory of instability of an inviscid fluid, because such theories are *structurally unstable* – that is, a small change in the model equations can make a substantial change in the results. For example, if a basic parallel flow of an inviscid incompressible fluid in Rossby's beta-plane has no normal mode growing with time (and therefore might be at once deemed stable), then a wave perturbation of small amplitude ϵ may be shown by weakly nonlinear theory to generate a perturbation of order of magnitude ϵ^{-1} in a thin critical layer of width ϵ after a long time of order ϵ^{-1} (Warn & Warn, 1978; Brown & Stewartson, 1979), and thereby render the flow unstable (at least according to the definition of stability for some norms).

For a typical basic flow, the only way to find many of the *stability characteristics*, that is, the properties of the eigenvalues s and eigenfunctions for all values of the Reynolds number, is to resort to computation. This makes it impracticable in a textbook to do more than show some of the numerical results and relate them to observations. However, for certain specially simple flows, more of the theory can be developed analytically. If $\mathbf{U}$, P are independent of one or more space variables, for example when the basic flow is invariant under translation or rotation, then we may separate one or more of the space variables. (The separation of the space variables may be justified by use of Fourier, or generalized Fourier, transforms.) Thus if $\mathbf{U}$ is independent of x and t, we may take normal modes of the form $\mathbf{u}'(\mathbf{x}, t) = e^{st+ikx}\hat{\mathbf{u}}(y, z)$, where k is the x-wavenumber; if $\mathbf{U}$ depends only on the cylindrical polar coordinate r, then we may take $\mathbf{u}' = e^{st+i(kx+n\theta)}\hat{\mathbf{u}}(r)$, where n is an (integral) azimuthal wavenumber and θ is the azimuthal angle about the x-axis. (It is understood here that each physical quantity is the real part of its complex representation, this being permissible because the problem is linear and homogeneous.) This may lead to an ordinary-differential, rather than a partial-differential, eigenvalue problem, and thus to a more tractable mathematical problem to solve in order to find the stability characteristics of the basic flow. It is for this mathematical reason, rather than any physical reason, that simple symmetric steady flows are treated so often in lecture courses, books and papers on hydrodynamic stability.

Example 2.11: The stability of Poiseuille pipe flow. If $\mathbf{U} = (1 - r^2)\mathbf{i}$, $P = p_0 - 4x/R$ and $\mathbf{u}'$, $p' \propto e^{st+i(kx+n\theta)}$, then the linearized stability problem (2.25)–(2.27) can be shown at length (see Exercise 2.17) to become

$$(s + ikU)\hat{u}_x + \frac{dU}{dr}\hat{u}_r = -ik\hat{p} + R^{-1}\left[\frac{d^2}{dr^2} + \frac{1}{r}\frac{d}{dr} - \left(k^2 + \frac{n^2}{r^2}\right)\right]\hat{u}_x,$$

$$(s + ikU)\hat{u}_r = -\frac{d\hat{p}}{dr} + R^{-1}\left[\frac{d^2}{dr^2} + \frac{1}{r}\frac{d}{dr} - \left(k^2 + \frac{1+n^2}{r^2}\right)\right]\hat{u}_r + \frac{2in}{Rr^2}\hat{u}_\theta,$$

$$(s + ikU)\hat{u}_\theta + \frac{U}{r}\hat{u}_\theta = -\frac{in}{r}\hat{p} + R^{-1}\left[\frac{d^2}{dr^2} + \frac{1}{r}\frac{d}{dr}\right.$$
$$\left. - \left(k^2 + \frac{1+n^2}{r^2}\right)\right]\hat{u}_\theta - \frac{2in}{Rr^2}\hat{u}_r,$$

$$ik\hat{u}_x + \frac{1}{r}\frac{d(r\hat{u}_r)}{dr} + \frac{in}{r}\hat{u}_\theta = 0,$$

where

$$\hat{u}_x = \hat{u}_r = \hat{u}_\theta = 0 \quad \text{at } r = 1$$

and

$$\hat{\mathbf{u}}, \hat{p} \text{ are nonsingular at } r = 0.$$

Here $U(r) = 1 - r^2$, $\hat{\mathbf{u}} = (\hat{u}_x, \hat{u}_r, \hat{u}_\theta)$ in cylindrical polar coordinates. This is a complicated, albeit a linear ordinary-differential, eigenvalue problem to determine s and the mode structure. Extensive numerical calculations (see Salwen *et al.*, 1980) indicate that, strangely enough, Re$(s) < 0$ for all modes, all R and all wavenumbers k, n. This implies that Poiseuille pipe flow is stable, in apparent contradiction to the experiments of Reynolds (1883); it is believed that, although the flow is stable to the infinitesimal perturbations of this theory, it is in practice unstable to perturbations of quite small finite amplitudes because of some form of subcritical instability. □

As a postscript to this chapter, recall §1.1 where it was noted that flow in a pipe appears to be stable to all small enough perturbations at any value of the Reynolds number R, but stable to all perturbations of all magnitudes only if $R < R_g$, say, where $R_g \approx 2000$. This is represented mathematically by defining a basic flow to be *globally asymptotically stable* if *any* perturbation vanishes after a long time, that is, relation (2.20) holds whatever the initial perturbation is; and defining R_g as the greatest number such that the flow is globally asymptotically stable for all $R < R_g$.

Exercises

2.1 *Turning point and transcritical bifurcation.* (a) Find the *explicit* general solution u of equation (2.7) of Example 2.6, and hence verify the sketch of the bifurcation diagram in Figure 2.9 for $l > 0$.

(b) Similarly, find the explicit solution u of equation (2.8) of Example 2.7, and verify the bifurcation diagram of Figure 2.10 for $l > 0$.

2.2 *The Landau equation.* Show that if u satisfies the equation

$$\frac{du}{dt} = k(R - R_c)u - lu^3$$

for k, R, $R_c > 0$ and real l, then

$$u^2(t) = \frac{k(R - R_c)u_0^2}{lu_0^2 + \left[k(R - R_c) - lu_0^2\right]\exp[-2k(R - R_c)t]}$$

so long as u remains finite. Deduce that the null solution $U(t) = 0$ is stable if $R < R_c$ and unstable if $R > R_c$. Verify the sketch of the bifurcation diagram in the (R, u)-plane for the cases $l > 0$, $l < 0$.

2.3   *The invariance of the complex Landau equation.* (i) Show that if the complex amplitude A satisfies the Landau equation,

$$\frac{\mathrm{d}A}{\mathrm{d}t} = sA - l|A|^2 A, \tag{E2.1}$$

for complex constants s, l, and if $B(T) = \mathrm{e}^{\mathrm{i}\theta} A(t), T = t + t_0$ for real constants θ, t_0, then

$$\frac{\mathrm{d}B}{\mathrm{d}T} = sB - l|B|^2 B.$$

Deduce that if $A(t)$ is a solution of equation (E2.1), then so is $\mathrm{e}^{\mathrm{i}\theta} A(t - t_0)$.
(ii) Show that if

$$\frac{\partial u}{\partial t} = a\frac{\partial^2 u}{\partial x^2} - b|u|^2 u \tag{E2.2}$$

for complex constants a, b, and we define $T = t - t_0$, $X = x - x_0$, then

$$\frac{\partial u}{\partial T} = a\frac{\partial^2 u}{\partial X^2} - b|u|^2 u,$$

so the equation is invariant under translations of time and space. Deduce that if $u(x, t) = A(t)\mathrm{e}^{\mathrm{i}kx}$ is a solution of equation (E2.2) for real k, then $u(x, t) = A(t - t_0)\mathrm{e}^{\mathrm{i}(kx+\theta)}$, where $\theta = -kx_0$, is another. Show, in particular, that if $t_0 = 0$, $x_0 = \pi/k$, then the sign of A is changed by the transformation.

2.4   *Bifurcation from infinity.* Given that

$$\frac{\mathrm{d}u}{\mathrm{d}t} = -u(1 - Ru)(1 - u)$$

for $R > 0$, show that $U = 0$ is a stable steady solution for all R. What other steady solutions are there? For what values of R are they stable? Sketch the bifurcation diagram.

  If $u = A$ at $t = 0$, for what values of A does $u(t) \to 0$ as $t \to \infty$? Comment on the significance of your answers when R is large.

2.5   *Secondary instability.* Given that

$$\frac{\mathrm{d}u}{\mathrm{d}t} = -u(1 - u)\left(1 - R + u^2\right),$$

show that the primary solution $U = 0$ has a pitchfork bifurcation at $R = 1$ but the secondary solution $U = (R - 1)^{1/2}$ has a secondary instability at a transcritical bifurcation where $R = 2$, $u = 1$.

2.6 *Different definitions of stability.* Consider the simple dynamical system

$$\frac{dx}{dt} = -y(x^2 + y^2)^{1/2}, \qquad \frac{dy}{dt} = x(x^2 + y^2)^{1/2}.$$

First show that the general solution is

$$x = a\cos(at + c), \qquad y = a\sin(at + c),$$

for arbitrary constants $a \geq 0$, c. Show further that of these solutions the only stable one is $x = y = 0$, on defining the norm $\|(x, y)\| = (x^2 + y^2)^{1/2}$.

Next, defining the new coordinates $r = (x^2 + y^2)^{1/2}$ and ϕ such that

$$x = r\cos(rt + \phi), \qquad y = r\sin(rt + \phi),$$

show that the above system is equivalent to

$$\frac{dr}{dt} = 0, \qquad \frac{d\phi}{dt} = 0.$$

Show that the general solution of this system is $r = a$, $\phi = c$, but that all of these solutions are stable, on defining the norm $\|(r, \phi)\| = (r^2 + \phi^2)^{1/2}$.

What is the meaning of these facts that the 'same' solution is stable in one set of coordinates but unstable in the other? [Cesari (1959, §1.9).]

2.7 *Mathematical and physical concepts of stability – let mathematics be your servant, not your master.* Consider the 'toy' mathematical system

$$\frac{\partial u}{\partial t} = x\frac{\partial u}{\partial x},$$

where $u(x, t) \to 0$ exponentially as $x \to \pm\infty$ for all $t \geq 0$.

Show that $u = U$ is the steady solution of this system, where $U(x) = 0$ for all x.

Show that if $u(x, 0) = \phi(x)$, where ϕ is non-null, infinitely differentiable, and decays exponentially at infinity, then $u(x, t) = \phi(xe^t)$ for all $t > 0$.

State the axiomatic properties a norm has to satisfy. Consider the definition

$$\|f\|_p = \left[\int_{-\infty}^{\infty}\left(\left|\frac{\partial f}{\partial x}\right|^p + |f|^p\right)dx\right]^{1/p},$$

for $p > 0$, and verify that it is a norm over an appropriate linear vector space. [Hint: use Minkowski's inequality.]

Show that, for the solution above,

$$\|u\|_p \sim e^{(1-1/p)t} \left[\int_{-\infty}^{\infty} |\phi'(y)|^p \, dy \right]^{1/p} \quad \text{as } t \to \infty.$$

Deduce that the solution U is asymptotically stable if $p < 1$ but exponentially unstable if $p > 1$.

Does this suggest that a given solution of some hydrodynamic problem may be both stable and unstable? Discuss. [Yudovich (1989, p. 101).]

2.8 *Incomplete and complete sets of normal modes.* (i) Suppose that

$$\frac{dx}{dt} = \mathbf{Ax}, \qquad \mathbf{x}(0) = [x_0, y_0]^{\mathrm{T}},$$

where

$$\mathbf{x} = \begin{bmatrix} x \\ y \end{bmatrix}, \quad \mathbf{A} = \begin{bmatrix} 0 & 1 \\ 0 & 0 \end{bmatrix}.$$

First show that

$$\mathbf{x}(t) = [x_0 + y_0 t, y_0]^{\mathrm{T}}.$$

Is the null solution stable?

Next seek to solve the problem again by the method of normal modes, taking $\mathbf{x}(t) \propto e^{st}$ and finding the eigenvalues s_k and associated independent eigenvectors $\mathbf{u}_k$ of $\mathbf{A}$. How many eigenvalues and eigenvectors are there? Can the method be used to solve the initial-value problem for all x_0, y_0?

(ii) Now use the method of normal modes to solve a small perturbation of the above problem, namely,

$$\frac{dx}{dt} = \mathbf{Bx}, \qquad \mathbf{x}(0) = [x_0, y_0]^{\mathrm{T}},$$

where

$$\mathbf{B} = \begin{bmatrix} -2\epsilon & 1 \\ 0 & -\epsilon \end{bmatrix} \quad \text{and} \quad 0 < \epsilon \ll 1.$$

Hence show that

$$\mathbf{x}(t) = (x_0 - y_0/\epsilon)\mathbf{u}_1 e^{-2\epsilon t} + (y_0/\epsilon)\mathbf{u}_2 e^{-\epsilon t},$$

where $\mathbf{u}_1 = [1, 0]^{\mathrm{T}}, \mathbf{u}_2 = [1, \epsilon]^{\mathrm{T}}$. [Note that the eigenvectors are nearly parallel.] How many eigenvalues and eigenvectors are there? Can the method be used to solve the initial-value problem for all x_0, y_0?

Deduce that $|\mathbf{x}(t)| \to 0$ as $t \to \infty$ for all fixed x_0, y_0 and ϵ. Is the null solution stable?

On the other hand, show that $\max_{t \geq 0} |\mathbf{x}(t)| \to \frac{1}{4}|y_0|/\epsilon$ as $\epsilon \to 0$ for all fixed x_0, $y_0 \neq 0$. Is the null solution stable? [Hint: it may help to define $E(t) = (1 - \epsilon x_0/y_0)e^{-\epsilon t}$ and find the turning points of $|\mathbf{x}(t)|^2 = [y_0/\epsilon(1 - \epsilon x_0/y_0)]^2 E(t)^2 \{[1 - E(t)]^2 + \epsilon^2\}$. For the relevance of this to hydrodynamic stability see Exercise 8.44, Trefethen *et al.* (1993) and Waleffe (1995).]

2.9 *The pressure gradient in Jeffery–Hamel flows.* Show that in the Jeffery–Hamel problem the radial pressure gradient is

$$\frac{\partial p}{\partial r} = \frac{\rho Q^2}{4r^3}\left[\left(\frac{\mathrm{d}\Psi}{\mathrm{d}\theta}\right)^2 + \frac{1}{R}\frac{\mathrm{d}^3\Psi}{\mathrm{d}\theta^3}\right],$$

where ρ is the density of the fluid.

Deduce that there is an adverse pressure gradient at a wall if $\mathrm{d}^2 U/\mathrm{d}\theta^2 < 0$ there. Show further that $[\mathrm{d}U/\mathrm{d}\theta]_{\theta=\alpha}$ changes sign from negative to positive for solutions of types II_1, II_2 as R increases through R_2.

2.10 *The exact general solution of the Jeffery–Hamel problem.* Show that an integral of the Jeffery–Hamel equation (2.5),

$$\Psi^{\mathrm{iv}} + 4\Psi'' + 2R\Psi'\Psi'' = 0,$$

is

$$\Psi''' + 4\Psi' + R(\Psi')^2 = A,$$

where $\Psi' = \mathrm{d}\Psi/\mathrm{d}\theta$ and A is a constant of integration. Deduce that

$$\frac{1}{2}(\Psi'')^2 + 2(\Psi')^2 + \frac{1}{3}R(\Psi')^3 = A\Psi' + B,$$

and thence indicate that the solutions Ψ of equation (2.5) and its boundary conditions (2.6),

$$\Psi(\theta) = \pm 1, \qquad \Psi'(\theta) = 0 \quad \text{at } \theta = \pm\alpha,$$

can be expressed in terms of Jacobian elliptic functions. [Jeffery (1915), Hamel (1916), Abramowitz & Stegun (1964, §17.4.61); see Fraenkel (1962), Batchelor (1967, §5.6).]

2.11 *Velocity profiles of types I, II_n and III_n for Stokes flow.* Show that if $R = 0$, then the Jeffery–Hamel problem (2.5), (2.6) above has solution

$$\Psi(\theta) = \frac{\sin 2\theta - 2\theta \cos 2\alpha}{\sin 2\alpha - 2\alpha \cos 2\alpha}.$$

2.12 *Plane Poiseuille flow.* Show that if $\alpha \to 0$ for fixed θ/α, then the Jeffery–Hamel problem (2.5), (2.6) has limiting solution $\Psi(\theta) = 3\theta/2\alpha - \theta^3/2\alpha^3$, that is, $U(\theta) \sim \frac{3}{2}(1 - \theta^2/\alpha^2)/\alpha r$.

2.13 *Linearization of the vorticity equation.* Using the basic vorticity equation (2.5), taking small perturbations of the form $\psi = \Psi + \psi'$, and linearizing the vorticity equation (2.4), show that

$$\frac{\partial \zeta'}{\partial t} + \frac{1}{r}\frac{d\Psi}{d\theta}\frac{\partial \zeta'}{\partial r} + \frac{2}{r^4}\frac{d^2\Psi}{d\theta^2}\frac{\partial \psi'}{\partial \theta} + \frac{1}{r^3}\frac{d^3\Psi}{d\theta^3}\frac{\partial \psi'}{\partial r} = R^{-1}\Delta\zeta', \quad \text{(E2.3)}$$

where $\zeta' = -\Delta\psi'$. Now, taking *steady* normal modes of the form $\psi' = r^\lambda\phi(\theta)$, deduce the *Dean equation*, namely, that

$$\phi^{iv} + \left[\lambda^2 + (\lambda - 2)^2\right]\phi'' + \lambda^2(\lambda - 2)^2\phi$$
$$= R\left[(\lambda - 2)\Psi'(\phi'' + \lambda^2\phi) - 2\Psi''\phi' - \lambda\Psi'''\phi\right], \quad \text{(E2.4)}$$

where

$$\phi(\theta) = \phi'(\theta) = 0 \quad \text{at } \theta = \pm\alpha. \quad \text{(E2.5)}$$

This constitutes the *Dean problem* to determine eigenvalues λ and eigenfunctions ϕ given Ψ, R and α.

   Why cannot both variables t, r be separated? [Dean (1934), Banks *et al.* (1988).]

2.14 *The Orr–Sommerfeld problem for steady perturbations of plane Poiseuille flow.* Show that $r^\lambda \sim$ constant $\times\, e^{ikx}$ as $\alpha \to 0$ for fixed $k = -i\alpha\lambda$, αr_0, and $x = (r - r_0)/\alpha r_0$. Deduce that equations (E2.4), (E2.5) become

$$\phi^{iv} - 2k^2\phi'' + k^4\phi = ikR\left[U(\phi'' - k^2\phi) - U''\phi\right],$$

$$\phi(y) = \phi'(y) = 0 \quad \text{at } y = \pm 1,$$

as $\alpha \to 0$ for fixed $y = \theta/\alpha$, where a prime now denotes differentiation with respect to y, and $\Psi'(\theta) \to U(y) = \frac{3}{2}(1 - y^2)$.

2.15 *Four exact eigensolutions.* Verify the identity

$$(\Psi^{iv} + 4\Psi'' + 2R\Psi'\Psi'')' = \Psi^v + 4\Psi''' + 2R\Psi'\Psi''' + 2R(\Psi'')^2.$$

Deduce that eigensolutions of the problem (E2.4), (E2.5) are given by
(i) $\lambda = 0, \phi = \Psi'$, (ii) $\lambda = 2, \phi = \Psi'$, (iii) $\lambda = -1, \phi = \cos\theta\Psi'$, and
(iv) $\lambda = -1, \phi = \sin\theta\Psi'$, provided that $\Psi'' = 0$ at $\theta = \pm\alpha$.

[Dean (1934) gave solutions (i), (ii). In fact the vorticity equation (2.4) is invariant under the group of rotations $r \mapsto r, \theta \mapsto \theta + \delta, \psi \mapsto \psi$ for all real δ, and therefore $\Psi(\theta + \delta) = \Psi(\theta) + \delta\Psi'(\theta) + O(\delta^2)$ as $\delta \to 0$ is also a solution of the *equation*: this gives $\psi' = \Psi'$, i.e. $\lambda = 0, \phi = \Psi'$. It is also invariant under the translations $x \mapsto x + \delta\cos\beta, y \mapsto y + \delta\sin\beta, \psi \mapsto \psi$, for which $\Psi(\theta - \delta r^{-1}\sin(\theta - \beta)) = \Psi(\theta) - \delta r^{-1}\sin(\theta - \beta)\Psi'(\theta) + O(\delta^2)$ as $\delta \to 0$; therefore $\lambda = -1, \phi = \sin(\theta - \beta)\Psi'(\theta)$.]

2.16 *The eigensolutions for Stokes flow.* Show that if $R = 0$, then an *even* eigenfunction of the problem (E2.4), (E2.5) is

$$\phi(\theta) = \frac{\cos\lambda\theta}{\cos\alpha\lambda} - \frac{\cos(\lambda - 2)\theta}{\cos\alpha(\lambda - 2)},$$

where λ is a zero of $V(-\lambda)$ and V is defined by

$$V(p) = (p + 1)\sin 2\alpha + \sin 2\alpha(p + 1).$$

Deduce that $2 - \lambda$ is also an eigenvalue.

Find similarly the odd eigenfunctions.

2.17 *Stability problem of Poiseuille pipe flow.* Express the Navier–Stokes equations in terms of cylindrical polar coordinates (x, r, θ), using dimensionless variables and a Reynolds number R.

Consider a basic flow of an incompressible viscous fluid with velocity $\mathbf{U} = U(r)\mathbf{i}$ and pressure $P = P(x)$, where $U(r) = 1 - r^2$, $P(x) = P_0 - 4x/R$ for $0 \le r \le 1$. Linearize the equations with small perturbations such that $\mathbf{u} = \mathbf{U} + \mathbf{u}', p = P + p'$. [See Example 2.11.]

2.18 *Instability due to linear resonance.* Suppose that

$$\frac{dx}{dt} = -i\omega_1 x + \epsilon p_1 y, \qquad \frac{dy}{dt} = \epsilon p_2 x - i\omega_2 y,$$

where $\omega_2 = \omega_1 + b\epsilon$ for real ω_1, ϵ, b. Show then that the normal modes with $x, y \propto e^{-i\omega t}$ have the dispersion relation

$$\omega = \omega_1 + \frac{1}{2}b\epsilon \pm \epsilon(b^2 - 4p_1 p_2)^{1/2}.$$

Deduce that the null solution $x = y = 0$ is stable if $\epsilon = 0$, but unstable for small ϵ when $4p_1 p_2 > b^2$.

[The resonance is called 'avoided crossing' by solid-state physicists. It shows that a weak linear coupling of two neutrally stable waves may render their basic state unstable. Drazin (1989) showed that this is a mechanism whereby buoyancy, usually reckoned as a stabilizing influence, may destabilize short waves for a stable basic parallel flow (see, e.g., Drazin & Reid 1981, equation (44.36)).]

2.19 *Normal modes which are exact nonlinear solutions.* Suppose that $\mathbf{u} = \mathbf{U}(\mathbf{x}, t)$, $p = P(\mathbf{x}, t)$ is a solution of the Navier–Stokes equations,

$$\frac{\partial \mathbf{u}}{\partial t} + \mathbf{u} \cdot \nabla \mathbf{u} = -\frac{1}{\rho} \nabla p + \nu \Delta \mathbf{u}, \qquad \nabla \cdot \mathbf{u} = 0,$$

governing unbounded flow of a uniform incompressible viscous fluid. Regarding this as a basic flow, and considering perturbations such that $\mathbf{u} = \mathbf{U} + \mathbf{u}'$, $p = P + p'$, where $\mathbf{u}' = f(\mathbf{x}, t)\hat{\mathbf{u}}(t)$ for some functions f, $\hat{\mathbf{u}}$ such that $\nabla \cdot \mathbf{u}' = 0$, show that

$$\mathbf{u}' \cdot \nabla \mathbf{u}' = \mathbf{0}$$

identically.

Deduce that if the linearized problem has a normal mode solution of the above form such that $f(\mathbf{x}, t) = g(\mathbf{k} \cdot \mathbf{x} - \omega t)$, and $\hat{\mathbf{u}} \propto e^{\sigma t}$ for real $\mathbf{k}, \omega, \sigma$, then the perturbation is also an exact nonlinear solution.

Reverting to consideration of the basic flow, show that if $\mathbf{U}(\mathbf{x}, t) = \mathbf{S}(t)\mathbf{x} + \mathbf{U}_0(t)$, an unsteady linear shear flow for some 3×3 matrix function $\mathbf{S}$, then

$$\frac{d\mathbf{S}}{dt} + \mathbf{S}^2 = \mathbf{M}, \qquad \text{trace}(\mathbf{S}) = 0,$$

where $\mathbf{M}(t)$ is a symmetric matrix function. [Craik & Criminale (1986).]

2.20 *An instability of an elliptical flow.* Consider the steady basic flow of a uniform incompressible viscous fluid with

$$\mathbf{U} = \frac{2\Omega}{a^2 + b^2}\left(-a^2 y, b^2 x, 0\right), \qquad P = \frac{2a^2 b^2 \rho \Omega^2 \left(x^2 + y^2\right)}{\left(a^2 + b^2\right)^2}$$

for positive constants a, b, ρ, Ω. Show that the vorticity $\nabla \times \mathbf{U} = 2\Omega \mathbf{k}$, a constant. Verify that this flow satisfies the Navier–Stokes equations exactly, and that $\mathbf{U} \cdot \mathbf{n} = 0$ on the ellipsoid with equation

$$\frac{x^2}{a^2} + \frac{y^2}{b^2} + \frac{z^2}{c^2} = 1,$$

where $\mathbf{n}$ is a vector normal to the ellipsoid (although the flow does not satisfy the no-slip condition on the ellipsoid).

Consider velocity perturbations of the form

$$\mathbf{u}' = (aqz/c - ary/b, brx/a - bpz/c, cpy/b - cqx/a)$$

for some functions p, q, r of t alone. Deduce that the vorticity of the perturbation is

$$\nabla \times \mathbf{u}' = \left((b^2 + c^2)p/bc, (c^2 + a^2)q/ca, (a^2 + b^2)r/ab\right),$$

a constant vector at each instant, and thence, by linearization of the vorticity equation for an incompressible viscous fluid, that

$$(b^2 + c^2)\frac{dp}{dt} = \frac{2\Omega ab(b^2 - c^2)q}{a^2 + b^2},$$

$$(c^2 + a^2)\frac{dq}{dt} = \frac{2\Omega ab(c^2 - a^2)p}{a^2 + b^2},$$

$$\frac{dr}{dt} = 0.$$

Taking normal modes with $p, q \propto e^{st}$, show that

$$s^2 = \frac{4\Omega^2 a^2 b^2 (c^2 - b^2)(a^2 - c^2)}{(b^2 + c^2)(a^2 + c^2)(a^2 + b^2)^2},$$

and thence that the flow is unstable if $a < c < b$ or $b < c < a$ but stable (to these perturbations) otherwise.

[Lamb (1932, §384). Note that the stability found above for the case $b = a$ agrees with the stability of the basic flow of a rigid-body rotation (see Exercise 5.12), although there is instability if $0 < b - a \ll 1$ provided that $a < c < b$. Note that the value of the viscosity does not affect the results because the viscous terms vanish identically *for the special perturbations considered*.]

2.21 *Kelvin's principle of minimum energy and the stability of potential flows.* Prove that, of all flows of an incompressible inviscid fluid in a simply connected domain with prescribed normal flux at the surface, the irrotational flow has the least kinetic energy.

Discuss why a basic irrotational flow, although the flow of minimum kinetic energy, is not necessarily stable, by considering two-dimensional perturbations of the uniform basic flow $\mathbf{U} = U\mathbf{i}$ in the square $-L < x, y < L$. Show that the linearized vorticity-perturbation equation is satisfied by $\zeta'(x, y, t) = f(x - Ut, y)$ for any differentiable function f, and hence that the problem for the velocity perturbation

$\mathbf{u}'(x, y, t)$ is not well posed if the boundary conditions are merely that the normal component of $\mathbf{u}'$ vanishes on the sides of the square. If further $\zeta'(x, y, 0) = g(x, y)$ in the square, and $\zeta'(-L, y, t) = h(y, t)$ for $t > 0$, where $U > 0$ and g and h are given functions, find ζ' inside the square for $t > 0$. Deduce that ζ' need not be small as $t \to \infty$, even if $g(x, y) = 0$ for all x, y.

[After Drazin & Reid (1981, Problem 1.1). See Lamb (1932, §45). Hint: $\partial\zeta'/\partial t + U\partial\zeta'/\partial x = 0$, $\zeta'(x, y, t) = g(x - Ut, y)$ for $0 < t < (x + L)/U$, and $\zeta' = h(y, t - xU)$ for $(x + L)/U < t$.]

2.22 *Jeans instability of self-gravitating gas.* The dynamics of self-gravitating intergalactic gas are modelled by Poisson's equation,

$$\Delta\phi = 4\pi G\rho,$$

as well as Euler's equations, the continuity equation,

$$\rho\left(\frac{\partial\mathbf{u}}{\partial t} + \mathbf{u}\cdot\nabla\mathbf{u}\right) = -\nabla p - \rho\nabla\phi, \qquad \frac{\partial\rho}{\partial t} + \nabla\cdot(\rho\mathbf{u}) = 0,$$

and an equation of state, $p = p(\rho)$, for a barotropic inviscid fluid, where ϕ is the gravitational potential and G the gravitational constant.

Considering small perturbations of a basic state of rest with $\phi = \Phi(\mathbf{x})$, $\rho = R(\mathbf{x})$, $\mathbf{u} = \mathbf{0}$ and denoting them by primes such that

$$\phi = \Phi + \phi', \qquad \rho = R + r', \qquad \mathbf{u} = \mathbf{u}',$$

and linearizing the equations of motion by neglecting products of the perturbed quantities, deduce that

$$\Delta\phi' = 4\pi G\rho',$$

$$R\frac{\partial\mathbf{u}'}{\partial t} = -c^2\nabla\rho' - \rho'\nabla\Phi - R\nabla\phi', \qquad \frac{\partial\rho'}{\partial t} + \nabla\cdot(R\mathbf{u}') = 0,$$

where $c = [\mathrm{d}p(R)/\mathrm{d}R]^{1/2}$ is the basic velocity of sound.

Assuming that the basic state is uniform with $R = $ constant, $\Phi = $ constant (although the latter implies the unphysical result that $R = 0$ and we later assume the physical property that $R > 0$), and taking normal modes with $\mathbf{u}', \rho', \phi' \propto e^{i\mathbf{k}\cdot\mathbf{x}+st}$, show that

$$s^2 = 4\pi GR - c^2\mathbf{k}^2.$$

Deduce that a uniform gas is gravitationally unstable to long waves. [Jeans (1902).]

3

Kelvin–Helmholtz Instability

They that go down to the sea in ships, that do business in great waters;
These see the works of the Lord, and his wonders in the deep. For
He commandeth and raiseth the stormy wind, which lifteth up the
waves thereof.

Psalm cvii 23–25

To understand better the mechanisms and concepts of the linear theory of stability described in Chapter 2, it helps to follow some simple worked examples. You will discern certain features common to the examples: (1) identification of the physical mechanism of instability of a given flow and modelling of the instability by choice of an appropriate system of equations and boundary conditions; (2) choice of a solution satisfying the system to represent the basic flow; (3) linearization of the system for small perturbations of the chosen basic flow; (4) use of the method of normal modes; and (5) application of the results to understand or control the observed instability. First, we shall work through a classic problem that does not demand a lot of mathematics. It substantiates the important physical mechanism whereby a basic vorticity gradient tends to destabilize a flow.

3.1 Basic Flow

Consider the basic flow of two incompressible inviscid fluids in horizontal parallel infinite streams of different velocities and densities, one stream above the other. Then the basic flow is given by velocity, density and pressure:

$$\mathbf{U}(z) = \begin{cases} U_2\mathbf{i} \\ U_1\mathbf{i}, \end{cases} \quad \bar{\rho}(z) = \begin{cases} \rho_2 \\ \rho_1, \end{cases} \quad P(z) = \begin{cases} p_0 - g\rho_2 z & \text{for } z > 0 \\ p_0 - g\rho_1 z & \text{for } z < 0, \end{cases} \quad (3.1)$$

respectively, say, where U_1, U_2 are the velocities of the two streams, ρ_1, ρ_2 are the densities, p_0 is a constant pressure, z is the height and g is the acceleration due to gravity.

3.2 Physical Description of the Instability

Helmholtz (1868) remarked that 'every perfect geometrically sharp edge by which a fluid flows must tear it as under and establish a surface of separation, however slowly the rest of the fluid may move', thereby implicitly recognizing

45

the basic flow above, but the mathematical problem of instability was first posed and solved by Kelvin (1871); it is now called *Kelvin–Helmholtz instability*. Kelvin's motivation for choosing the basic flow above is not entirely clear, but he was a friend of Helmholtz and, as we shall see, applied the model to seek to understand the generation of ocean waves by the wind.

The physical mechanism of Kelvin–Helmholtz instability has been described by Batchelor (1967, pp. 515–516) in terms of the vorticity dynamics. For simplicity of description, let us suppose for the moment that $U_1 = -V$, $U_2 = V > 0$, and that $\rho_2 = \rho_1$, so that we consider the special case of a vortex sheet in a homogeneous fluid. Thus buoyancy is ignored here. Next consider an initial disturbance which slightly displaces the sheet so that its elevation is sinusoidal. Again for simplicity we suppose that the flow is two-dimensional in the (x, z)-plane, so that the elevation of the sheet is given by $z = \zeta(x, t)$ at subsequent times. Batchelor traced the vorticity dynamics as follows, using the fundamental properties that each vortex line in an inviscid fluid is carried with the fluid and induces a rotating flow with circulation equal to the strength of the vortex line. The vorticity $\partial u/\partial z - \partial w/\partial x$ of the sheet is positive for $V > 0$. So positive vorticity is swept away from points like A (in Figure 3.1) where $\zeta = 0, \partial\zeta/\partial x > 0$ and towards points like C where $\zeta = 0, \partial\zeta/\partial x > 0$, because vorticity in parts of the sheet displaced downwards (or upwards) induces a velocity with a positive (or negative) x-component at any part of the sheet where $z > 0$ (or <0). In particular, the induced velocity at points like B due to each part of the displaced vortex sheet has positive x-component. Now the positive velocity accumulating at points like C will induce clockwise

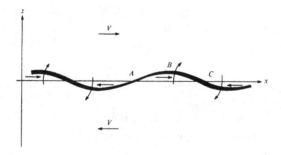

Figure 3.1   Growth of a sinusoidal disturbance of a vortex sheet with positive vorticity normal to the paper. The local strength of the sheet is represented by the thickness of the sheet. The arrows indicate the directions of the self-induced movement of the vorticity in the sheet, and show (a) the accumulation of vorticity at points like A and (b) the general rotation about points like A, which together lead to exponential growth of the disturbance. (After Batchelor, 1967, Fig. 7.1.3.)

velocities around such points and thereby amplify the sinusoidal displacement of the vortex sheet. These processes of accumulation of vorticity at points like C and of rotation of neighbouring points of the sheet will continue together, leading to exponential growth of the disturbance without change of the spatial form of the disturbance so long as the disturbance is small enough not to significantly change the basic state. We shall next substantiate this physical description with some mathematical details.

3.3 Governing Equations for Perturbations

Kelvin assumed that the disturbed flow was irrotational on each side of the vortex sheet. This follows if the initial disturbance of the flow is irrotational, because irrotational flow of an inviscid fluid persists. However, initial rotational disturbances also are possible. To simplify the mathematics we shall adopt Kelvin's restrictive assumption, remembering that it allows a proof of instability but not stability because it gives no information about rotational disturbances. In fact, as we shall see in Exercise 3.4 and Chapter 8, rotational disturbances are no more unstable, and so Kelvin did find a necessary as well as a sufficient condition for instability. Thus we assume the existence of a velocity potential ϕ on each side of the interface between the two streams with $\mathbf{u} = \nabla\phi$, where

$$\phi = \begin{cases} \phi_2 & \text{for } z > \zeta \\ \phi_1 & \text{for } z < \zeta, \end{cases} \tag{3.2}$$

the interface having elevation

$$z = \zeta(x, y, t) \tag{3.3}$$

when the flow is disturbed. Then the equations of continuity and incompressibility give $\nabla \cdot \mathbf{u} = 0$ and therefore the Laplacians of the potentials vanish, i.e.

$$\Delta\phi_2 = 0 \quad \text{for } z > \zeta, \qquad \Delta\phi_1 = 0 \quad \text{for } z < \zeta. \tag{3.4}$$

Note that Euler's equations of motion have been used only implicitly in taking the irrotational flow as persistent.

The boundary conditions are as follows:

(a) The initial disturbance may be supposed to occur in a finite region so that for all time

$$\nabla\phi \to \mathbf{U} \quad \text{as } z \to \pm\infty. \tag{3.5}$$

(b) The fluid particles at the interface must move with the interface without the two fluids occupying the same point at the same time and without a cavity forming between the fluids. Therefore the vertical velocity at the interface is given by

$$\frac{\partial \phi}{\partial z} = \frac{D\zeta}{Dt} = \frac{\partial \zeta}{\partial t} + \frac{\partial \phi}{\partial x}\frac{\partial \zeta}{\partial x} + \frac{\partial \phi}{\partial y}\frac{\partial \zeta}{\partial y} \quad \text{at } z = \zeta, \tag{3.6}$$

the material derivative of the surface elevation (see Lamb, 1932, p. 7). This kinematic condition is the same as that for surface gravity waves, which will be shown to be a special form of this instability. There is a discontinuity of tangential velocity at the interface, and it leads to the two equations,

$$\frac{\partial \phi_k}{\partial z} = \frac{\partial \zeta}{\partial t} + \frac{\partial \phi_k}{\partial x}\frac{\partial \zeta}{\partial x} + \frac{\partial \phi_k}{\partial y}\frac{\partial \zeta}{\partial y} \quad \text{at } z = \zeta, \text{ for } k = 1, 2. \tag{3.7}$$

(c) The normal stress of the fluid is continuous at the interface. For an inviscid fluid, this gives the dynamical condition that the pressure is continuous. Therefore

$$\rho_1 \left[C_1 - \tfrac{1}{2}(\nabla\phi_1)^2 - \partial\phi_1/\partial t - gz \right]$$
$$= \rho_2 \left[C_2 - \tfrac{1}{2}(\nabla\phi_2)^2 - \partial\phi_2/\partial t - gz \right] \quad \text{at } z = \zeta, \tag{3.8}$$

by Bernoulli's theorem for irrotational flow, which is valid on each side of the vortex sheet, $z = \zeta$. In order that the basic flow satisfies this condition, the constants C_1, C_2 must be related so that

$$\rho_1 \left(C_1 - \tfrac{1}{2}U_1^2 \right) = \rho_2 \left(C_2 - \tfrac{1}{2}U_2^2 \right). \tag{3.9}$$

Equations (3.2)–(3.9) pose the full nonlinear problem for perturbations of the basic flow (3.1).

3.4 The Linearized Problem

For linear stability we first express

$$\phi_2 = U_2 x + \phi_2' \quad \text{for } z > \zeta, \qquad \phi_1 = U_1 x + \phi_1' \quad \text{for } z < \zeta \tag{3.10}$$

and neglect products of the small perturbations ϕ_1', ϕ_2', ζ. This is effectively a definition of the perturbed quantities. There is no length scale in the basic velocity, so it is far from clear how small ζ must be in order that the linearization is valid. However, we can plausibly justify the linearization if the surface displacement and its slopes are small in dimensionless terms, i.e. $\partial\zeta/\partial x, \partial\zeta/\partial y \ll 1$ and $g\zeta \ll U_1^2, U_2^2$. If this is granted, and it is recognized that the exact position of the interface at $z = \zeta$ may be replaced by the basic position at $z = 0$ on using the Taylor series

$$[\phi'_k]_{z=\zeta} = [\phi'_k]_{z=0} + \zeta[\partial\phi'_k/\partial z]_{z=0} + \cdots,$$

then linearization of equations (3.4)–(3.8) is straightforward, giving

$$\Delta\phi'_2 = 0 \text{ for } z > 0, \qquad \Delta\phi'_1 = 0 \quad \text{for } z < 0; \qquad (3.11)$$

$$\nabla\phi'_k \to \mathbf{0} \quad \text{as } z \to \mp\infty \text{ for } k = 1, 2 \text{ respectively}; \qquad (3.12)$$

$$\frac{\partial\phi'_k}{\partial z} = \frac{\partial\zeta}{\partial t} + U_k\frac{\partial\zeta}{\partial x} \quad \text{at } z = 0, \text{ for } k = 1, 2. \qquad (3.13)$$

$$\rho_1\left(U_1\partial\phi'_1/\partial x + \partial\phi'_1/\partial t + g\zeta\right) = \rho_2\left(U_2\partial\phi'_2/\partial x + \partial\phi'_2/\partial t + g\zeta\right) \quad \text{at } z = 0. \qquad (3.14)$$

It can be seen that all coefficients of this linear partial differential system are constants and that the boundaries are horizontal. So we use the method of normal modes, assuming that an arbitrary disturbance may be resolved into independent modes of the form

$$(\zeta, \phi'_1, \phi'_2) = (\hat{\zeta}, \hat{\phi}_1(z), \hat{\phi}_2(z))e^{i(kx+ly)+st}. \qquad (3.15)$$

This reduces the problem to an ordinary differential system with z as the dependent variable, for here $\hat{\phi}_1, \hat{\phi}_2$ are functions of z only and $\hat{\zeta}$ is a constant.

Equations (3.11) now give

$$\hat{\phi}_2(z) = A_2e^{-\tilde{k}z} + B_2e^{\tilde{k}z}, \qquad (3.16)$$

where A_2 and B_2 are arbitrary constants and $\tilde{k} = (k^2 + l^2)^{1/2}$ is the total wavenumber. The boundary condition (3.12) at plus infinity implies that $B_2 = 0$ and therefore

$$\hat{\phi}_2(z) = A_2e^{-\tilde{k}z}. \qquad (3.17)$$

Similarly, we find that

$$\hat{\phi}_1(z) = A_1e^{\tilde{k}z}. \qquad (3.18)$$

Now equations (3.13), (3.14) give three homogeneous linear algebraic equations for the three unknown constants $\hat{\zeta}$, A_1, A_2. Equations (3.13) give

$$A_2 = -(s + ikU_2)\hat{\zeta}/\tilde{k}, \qquad A_1 = (s + ikU_1)\hat{\zeta}/\tilde{k} \qquad (3.19)$$

and thence give the eigenfunctions (3.15) except for an arbitrary multiplicative constant. Then equation (3.14) gives the eigenvalue relation,

$$\rho_1\big[\tilde{k}g + (s + ikU_1)^2\big] = \rho_2\big[\tilde{k}g - (s + ikU_2)^2\big]. \tag{3.20}$$

The solution of this quadratic equation gives two modes with

$$s = -ik\frac{\rho_1 U_1 + \rho_2 U_2}{\rho_1 + \rho_2} \pm \left[\frac{k^2\rho_1\rho_2(U_1 - U_2)^2}{(\rho_1 + \rho_2)^2} - \frac{\tilde{k}g(\rho_1 - \rho_2)}{\rho_1 + \rho_2}\right]^{1/2}. \tag{3.21}$$

Both are neutrally stable if

$$\tilde{k}g\big(\rho_1^2 - \rho_2^2\big) \geq k^2\rho_1\rho_2(U_1 - U_2)^2, \tag{3.22}$$

the equality giving marginal stability. One mode is asymptotically stable but the other unstable if

$$\tilde{k}g\big(\rho_1^2 - \rho_2^2\big) < k^2\rho_1\rho_2(U_1 - U_2)^2. \tag{3.23}$$

This is accordingly a necessary and sufficient condition for instability of the mode with wavenumbers k, l. Thus the flow is always unstable (to modes with sufficiently large k, that is, to short waves) if $U_1 \neq U_2$.

3.5 Surface Gravity Waves

To interpret this result it is simplest to consider special cases separately. When $\rho_2 = 0$ and $U_1 = U_2 = 0$ we have the model of surface gravity waves on deep water. They are stable with phase velocity

$$c = is/\tilde{k} = \pm\big(g/\tilde{k}\big)^{1/2}, \tag{3.24}$$

as is well known. This illustrates the identity of waves and oscillatory stable normal modes. It is often helpful to regard waves as a special case of hydrodynamic stability.

3.6 Internal Gravity Waves

When the basic flow is at rest ($U_1 = U_2 = 0$) we find

$$s = \pm\big[\tilde{k}g(\rho_2 - \rho_1)/(\rho_1 + \rho_2)\big]^{1/2}. \tag{3.25}$$

There is instability if and only if $\rho_1 < \rho_2$, that is, heavy fluid rests above light fluid. However, if $\rho_1 > \rho_2$, then there is stability, a normal mode being a wave with phase velocity

$$c = \pm\left[g(\rho_1 - \rho_2)/\tilde{k}(\rho_1 + \rho_2)\right]^{1/2}. \tag{3.26}$$

The eigenfunctions (3.15) for the velocity potential die away exponentially with distance from the interface, as in all cases of Kelvin–Helmholtz instability, and thus the motion is confined to the vicinity of the interface between the two fluids. These waves are a special case of *internal gravity waves*, which may propagate in the interior of a stratified fluid far from any boundary. They can be observed between layers of fresh and salt water that occur in estuaries; the upper surface of the fresh water may be very smooth while strong internal gravity waves occur at the interface of the salt water a metre or two below, because for fresh and salt water $(\rho_1 - \rho_2)/(\rho_1 + \rho_2) \approx 10^{-2} \ll 1$, and so equations (3.19) give relatively small fluid velocities for given amplitude of interfacial elevation.

3.7 Rayleigh–Taylor Instability

The eigenvalues (3.25) can be interpreted differently if the whole fluid system has an upward vertical acceleration f. Then, by the principle of equivalence in dynamics, or by solution of the problem of normal modes,

$$s = \pm\left[\tilde{k}g'(\rho_2 - \rho_1)/(\rho_1 + \rho_2)\right]^{1/2}, \tag{3.27}$$

where $g' = f + g$ is the apparent gravitational, or net vertical, acceleration of the system. It follows that there is instability if and only if $g' < 0$, i.e. the net acceleration is directed from the lighter towards the heavier fluid. This is called *Rayleigh–Taylor instability*, after Rayleigh's (1883) theory of the stability of a stratified fluid at rest under the influence of gravity and Taylor's (1950) recognition of the significance of accelerations other than gravity (see Exercise 8.20).

Rayleigh–Taylor instability can be simply observed by rapidly accelerating a glass of water downwards (and standing clear!). Quantitative observations have been made by Lewis (1950) and others. There are many applications of the theory. A spectacular one arises in a *young* supernova, where gas concentrates in a thin shell behind an interface which decelerates after the initial explosion, so that dense gas lies inside less dense gas as the expanding sphere sweeps up interstellar gas (Gull, 1975). The consequent breaking of symmetry and formation of filaments is observable long afterwards (see Figure 3.2).

Figure 3.2   The residue of the early Rayleigh–Taylor instability of the Crab Nebula, visible long afterwards as the gas expands into space. (Reproduced by permission of the Chandra X-Ray Observatory.)

3.8 Instability Due to Shear

The essential mechanism of shear instability in the absence of buoyancy is of widespread importance. Lundgren (1982) suggested that this instability is fundamental to maintaining turbulence at high values of the Reynolds number by breaking up shear layers.

When there is a vortex sheet in a homogeneous fluid ($\rho_1 = \rho_2$, $U_1 \neq U_2$), equation (3.21) gives

$$s = -\tfrac{1}{2}ik(U_1 + U_2) \pm \tfrac{1}{2}k(U_1 - U_2). \qquad (3.28)$$

One of these modes grows exponentially, so the flow is always unstable; the modes are growing and decaying waves with the phase velocity $c = \tfrac{1}{2}k(U_1 + U_2)/\tilde{k}$, the average velocity of the basic flow resolved in the direction $(k/\tilde{k}, l/\tilde{k})$ of propagation. Waves of all lengths are unstable, there being no length scale of the basic flow. Further, the growth rate is proportional to k, so that short waves grow fastest but there is no fastest growing wave. It in fact follows that the linearized initial-value problem for this flow is ill-posed, and a singularity in the profile of the vortex sheet may develop in a finite time, even if the initial data are analytic. It can also be seen that the wave of a given length $\lambda = 2\pi/\tilde{k}$ which grows most rapidly is the one which propagates in the direction of the basic flow (that is, $k = \tilde{k}$). So, after some time, waves in the direction of the flow will become dominant.

It might have been anticipated that there would be no margin of stability for a vortex sheet of an inviscid fluid, because there is no dimensionless combination of the physical quantities $U_2 - U_1, \rho_1$ governing the growth of perturbations in *this* model. (Only the difference of velocities is relevant because we may, without loss of generality, assign any value we please to the mean velocity $\frac{1}{2}(U_1 + U_2)$ by making a Galilean transformation.) Taking the characteristic length scale L of the wave as the wavelength λ and the characteristic velocity scale as half the basic difference $V = \frac{1}{2}(U_1 - U_2)$, we find the characteristic time scale $L/V = 4\pi/k(U_1 - U_2)$, during which time the wave amplitude grows by a factor $\exp(kc_i L/V) = \exp(2\pi k/\tilde{k}) \approx \exp(2\pi) \approx$ 536. In practice the instability does not grow as rapidly as that, because of the idealizations in Kelvin's model: although equation (3.28) describes well the very rapid growth of long small-amplitude waves on a shear layer of a slightly viscous fluid, the model is over-simplified as a description of instability of a real fluid. However, the amplitude grows rapidly, and nonlinearity soon both skews the profile of the wave in the direction of the basic flow and rolls up the interface, as sketched in Figure 3.3.

For a real shear layer of finite thickness, called a *free shear layer* or a *mixing layer*, we shall show (in Chapter 8) that short waves are stable. Neu (1984) has described more realistically the roll-up of a vortex sheet in a slightly viscous fluid and the secondary instablities which develop. However, the above result does demonstrate the instability of a shear layer to waves whose lengths are much greater than the thickness of the layer. Also Kelvin (1871, see Exercise 3.2) himself incorporated surface tension as well as buoyancy into the model, and found a margin of stability of the vortex sheet marked by a critical value of a dimensionless parameter.

The full condition (3.23) of Kelvin–Helmholtz instability represents an imbalance between the destabilizing effect of inertia and the stabilizing effect of buoyancy when the heavier fluid is below. Kelvin (1871) used this theory as a model of the generation of ocean waves by wind. Helmholtz (1890) applied the theory to billow clouds, whose appearance in regular lines marks the instability of winds with strong shear. A striking photograph of billow clouds is shown in Figure 3.4(a), the photographer having the good fortune to view a regular line of clouds in a direction perpendicular to that of wave propagation. The billow clouds, when seen obliquely, are often call a 'mackerel sky' after

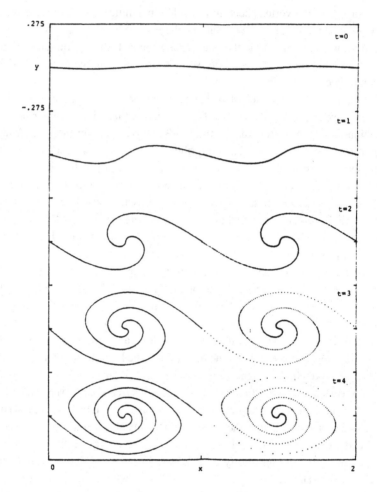

Figure 3.3   Sketch of the roll-up of a vortex sheet due to the strongly nonlinear growth of a two-dimensional perturbation. The interface has been computed at various time intervals. (After Krasny, 1986, Fig. 2; reproduced by permission of Academic Press.)

the barred pattern of the skin of the North Atlantic fish. The instability leads to what is called *clear-air turbulence* when there is insufficient water vapour to form clouds and thereby make the instability visible.

Reynolds's (1883) paper on pipe flow also describes some experiments on Kelvin–Helmholtz instability, although he did not recognize them as such. Reynolds filled a pipe with water above carbon disulphide and tilted it. The ensuing relative motion of the two fluids led to instability at their interface.

Thorpe (1969) later refined this experiment and clearly identified the Kelvin–Helmholtz instability. Some of his results are illustrated in Figure 3.4(b) and (c).

It must be emphasized that this crude model of Kelvin is a valuable first attempt to understand the instability, but it does not model many important features of the observed instability, such as the non-zero thickness of the shear layer, the effects of viscosity and the nonlinearity of the perturbation. Also Kelvin made no attempt to find the ultimate evolution of the instability. As you can see from Figure 3.4, in practice the vortex sheet steepens and then rolls up; Klaasen & Peltier (1985) showed theoretically and Thorpe (1985) experimentally that a secondary three-dimensional Rayleigh–Taylor instability in the billow cores develops later and thereafter the shear layer finally breaks up into turbulence. However, there is a lot more which might be written about this complicated phenomenon.

Surface waves are illustrated on film by Bryson (FL1967, containing excerpts from F1967) and internal gravity waves by Long (F1968, a few sequences only). Kelvin–Helmholtz instability modified by viscosity is shown in one

Figure 3.4   Kelvin–Helmholtz instability. (a) Billow clouds near Denver, Colorado, photographed by Paul E. Branstine. For the meteorological details, see Colson (1954).

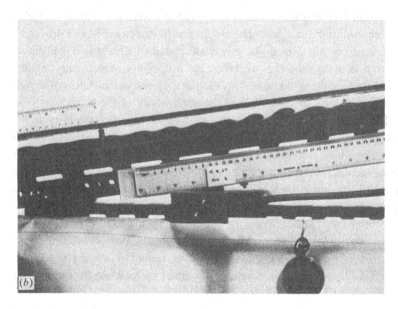

Figure 3.4 (*continued*) (b) Development of instability at the interface of two fluids of equal depth in relative acceleration owing to the tilt of the channel. (From Thorpe, 1968.)

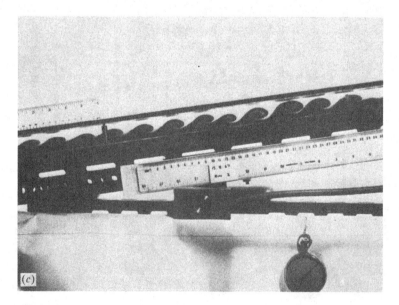

Figure 3.4 (*continued*) (c) The same run of Thorpe's experiment about half a second later.

short sequence by Mollo-Christensen & Wille (FL1968) and Rayleigh–Taylor instability modified by surface tension in another.

Exercises

3.1 *Laplace's law of surface tension.* Let S be a bounded simply connected smooth subsurface of the interface of a liquid and a gas with a smooth closed perimeter C. Assuming that the surface tension γ leads to the force $\gamma \int_C \mathbf{n} \times d\mathbf{x}$ on C and that this force is in equilibrium with the net force due to the pressure difference Δp across the interface, show that

$$\gamma \int_C \mathbf{n} \times d\mathbf{x} = \int \int_S (\Delta p)\mathbf{n} \, dS.$$

Deduce that

$$\Delta p = \gamma \nabla \cdot \mathbf{n}. \tag{E3.1}$$

3.2 *The moderation of Kelvin–Helmholtz instability by surface tension.* Show that if there is surface tension γ between the two fluids of §3.3, then equation (3.21) is replaced by

$$s = -ik\frac{\rho_1 U_1 + \rho_2 U_2}{\rho_1 + \rho_2} \pm \left\{ \frac{k^2 \rho_1 \rho_2 (U_1 - U_2)^2}{(\rho_1 + \rho_2)^2} \right.$$
$$\left. - \frac{\tilde{k}^2}{\rho_1 + \rho_2}\left[\frac{g(\rho_1 - \rho_2)}{\tilde{k}} + \tilde{k}\gamma\right]\right\}^{1/2}. \tag{E3.2}$$

Deduce that the flow is stable if and only if

$$(U_1 - U_2)^2 \leq 2(\rho_1 + \rho_2)[g\gamma(\rho_1 - \rho_2)]^{1/2}/\rho_1\rho_2, \tag{E3.3}$$

the wavelength of the least stable wave on the margin of stability being $\lambda = 2\pi/\tilde{k} = 2\pi[\gamma/g(\rho_1 - \rho_2)]^{1/2}$.

Using this model of Kelvin, show that the wind generates waves on the sea if the difference of the basic air and water speeds is such that

$$|U_1 - U_2| > 6.6 \, \text{m s}^{-1},$$

and the least stable wave has length $\lambda = 0.017$ m and speed 0.008 m s^{-1}.

[After Drazin & Reid (1981, Problem 1.4); Kelvin (1871). Hint: Use equation (E3.1) of Exercise 3.1 to incorporate the effect of surface tension on the dynamic condition at the interface. Hint: use $\rho_1 = 1020 \, \text{kg m}^{-3}$, $\rho_2 = 1.25 \, \text{kg m}^{-3}$, $g = 9.8 \, \text{m s}^{-2}$, $\gamma = 0.074 \, \text{N m}^{-1}$.]

3.3 *An initial-value problem for the perturbation of a vortex sheet.* Suppose that the basic flow is a vortex sheet in a homogeneous incompressible inviscid fluid, taking equation (3.2) with $U_2 = V > 0$, $U_1 = -V$, $\rho_2 = \rho_1$. Consider an irrotational two-dimensional perturbation for which the interface is released from rest with

$$\zeta(x, 0) = H \exp(-x^2/2L^2).$$

Deduce that

$$\zeta(x, t) = H \exp\left[(V^2 t^2 - x^2)/2L^2\right] \cos\left(V x t / L^2\right) \quad \text{for } t > 0,$$

finding ϕ_1', ϕ_2'.

[After Drazin & Reid (1981, Problem 1.5). Hint:

$$\zeta(x, 0) = (2\pi)^{-1/2} H L \int_{-\infty}^{\infty} \exp\left(-\tfrac{1}{2}k^2 L^2 + ikx\right) dk.$$

Therefore

$$\zeta(x, t) = (2\pi)^{-1/2} H L \int_{-\infty}^{\infty} \exp\left(-\tfrac{1}{2}k^2 L^2 + ikx\right) \cosh(kVt)\, dk,$$

because $s = \pm kV$ and $\partial \zeta / \partial t = 0$ at $t = 0$. Therefore

$$\phi_1'(x, z, t) = 2(2\pi)^{-1/2} H L V \int_{0}^{\infty} \exp\left(-\tfrac{1}{2}k^2 L^2 + ikz\right)$$
$$\times (\sin kx \cosh kVt + \cos kx \sinh kVt)\, dk,$$

with a similar expression for ϕ_2' differing only in some signs. Note that $\nabla \phi_1'$, $\nabla \phi_2' \neq \mathbf{0}$ at $t = 0$.]

3.4 *The effect of rotational disturbances on Kelvin–Helmholtz instability.* Show that, in the problem of §3.3, the linearized Euler equations of motion give

$$\frac{\partial \mathbf{u}'}{\partial t} + U_2 \frac{\partial \mathbf{u}'}{\partial x} = -\rho_2^{-1} \nabla p' \quad \text{for } z > \zeta.$$

Deduce that $\Delta p' = 0$ and that the perturbation of vorticity satisfies

$$\boldsymbol{\omega}'(\mathbf{x}, t) = \boldsymbol{\omega}'(x - U_2 t, y, z) \quad \text{for } z > \zeta.$$

Hence argue plausibly that the presence of vorticity in the initial perturbation does not affect the criterion of instability (3.23). [After Drazin & Reid (1981, Problem 1.6).]

3.5 *Rayleigh–Taylor stability of superposed fluids confined by a vertical cylinder.* Consider an incompressible inviscid fluid of density ρ_1 beneath a similar fluid of density ρ_2, both fluids being at rest and confined by a long vertical rigid cylinder of radius $r = a$, and there being surface tension γ at the horizontal interface $z = 0$. Here cylindrical polar coordinates r, θ, z are used, and Oz is the upward vertical. Then show, much as in Exercise 3.2, that small irrotational perturbations of the state of rest may be found as a superposition of normal modes of the form

$$\phi' \propto J_n(kr) \cos n\theta \, e^{-k|z|+st} \quad \text{and} \quad \zeta \propto J_n(kr) \cos n\theta \, e^{st}$$

for $n = 0, 1, \ldots$, where

$$s^2 = \frac{g(\rho_2 - \rho_1)k - \gamma k^3}{\rho_1 + \rho_2}$$

and $ka = j'_{n,m}$, the mth positive zero of the derivative J'_n of Bessel's function of nth order for $m = 1, 2, \ldots$. Deduce that there is stability if

$$a^2 g(\rho_2 - \rho_1) < \gamma j'^{2}_{1,1}.$$

[After Drazin & Reid (1981, Problem 1.12). This result shows that the stabilization of long waves by a boundary and of short waves by surface tension may stabilize heavy fluid resting above light fluid to all perturbations. This physical point is made visually by the experiment of Mollo-Christensen (F1968), for real, and therefore viscous, fluids for which short waves are further stabilized by viscosity. Maxwell (1876).]

3.6 **Rayleigh–Taylor instability with viscosity and surface tension.* Consider two incompressible fluids at rest, separated by a horizontal boundary, say $z = 0$. Let ρ_1 be the density of the lower fluid, and ρ_2 be the density of the upper. Suppose that both fluids have the same kinematic viscosity ν, and that there is surface tension γ at their interface. Taking small perturbations and linearizing, show that normal modes proportional to $e^{st+i(kx+ly)}$ are governed by the eigenvalue relation,

$$(\rho_1 + \rho_2)\big[g\tilde{k}(\rho_1 - \rho_2) + \tilde{k}^3 \gamma\big](r - 1) + 4\tilde{k}^2 s\nu(\rho_1 - \rho_2)^2 (r - 1)$$
$$+ 4s^2 \rho_1 \rho_2 = 4\tilde{k}^4 \nu^2 (\rho_1 - \rho_2)^2 (r - 1)^2,$$

where $\tilde{k}^2 = k^2 + l^2$, $r = (1 + s/\tilde{k}^2\nu)^{1/2}$ and $\mathrm{Re}(r) > 0$.

[Hints: Linearize the equations of motion, and then set $w' = \hat{w}_j(z) \times e^{st+i(kx+ly)}$, where $j = 1$ below the interface and $j = 2$ above it; deduce that the equations of motion and boundary conditions at infinity are satisfied by $\hat{w}_1(z) = A_1 e^{\tilde{k}z} + B_1 e^{\tilde{k}rz}$, $\hat{w}_2(z) = A_2 e^{-\tilde{k}z} + B_2 e^{-\tilde{k}rz}$. Show that the continuity of the normal and tangential components of the velocity at the disturbed interface gives continuity of $\hat{w}$, $D\hat{w}$ respectively at $z = 0$. Show that the continuity of the tangential and normal components of the stress at the disturbed interface gives $\rho_2(D^2 + \tilde{k}^2)\hat{w}_2 = \rho_1(D^2 + \tilde{k}^2)\hat{w}_1$, $\rho_2[1 - vs^{-1}(D^2 - \tilde{k}^2)]D\hat{w}_2 - \rho_1[1 - vs^{-1}(D^2 - \tilde{k}^2)]D\hat{w}_2 = -\tilde{k}^2 s^{-2}[g(\rho_2 - \rho_1) - \tilde{k}^2 \gamma]\hat{w} - 2\tilde{k}^2 vs^{-1}(\rho_2 - \rho_1)D\hat{w}$ respectively at $z = 0$.]

[Harrison (1908), Bellman & Pennington (1954), see, e.g., Chandrasekhar (1961, §94).]

3.7 *Rayleigh–Taylor instability and Schwarzschild's criterion for atmospheric stability.* An incompressible inviscid fluid of density $\bar{\rho}(z)$ at rest can be shown to be stable if $\bar{\rho}'(z) < 0$ for all z by the following heuristic 'physical' argument. 'Suppose that a fluid particle at level z_0 is raised (or lowered if $\delta z < 0$) a little to height $z_0 + \delta z$ somehow. Then its density remains $\bar{\rho}(z_0)$ (because it is incompressible and density is not diffused), but the ambient fluid in the particle's new environment has density $\bar{\rho}(z_0 + \delta z) = \bar{\rho}(z_0) + \delta z \bar{\rho}'(z_0) + O[(\delta z)^2]$ as $\delta z \to 0$. Therefore if the particle is denser than the ambient fluid (lighter if $\delta z < 0$), then $\bar{\rho}(z_0) > \bar{\rho}(z_0 + \delta z)$, i.e. $\bar{\rho}'(z_0) < 0$, and the particle is pulled back by the buoyancy towards its original level z_0 as if by a spring. Conversely, if $\bar{\rho}'(z_0) < 0$, then the particle continues to rise away from its original level. It follows that the fluid is stable if $\bar{\rho}'(z) < 0$ for all z and unstable if $\bar{\rho}'(z) > 0$ for some z.' [Rayleigh (1883) linearized the Euler equations to demonstrate this result more convincingly. See Exercise 8.19.]

You are given that an atmosphere of a perfect gas at rest is stable to small adiabatic perturbations if $\Gamma < \Gamma_a$ and unstable if $\Gamma > \Gamma_a$, where Γ is the *lapse rate* of the atmosphere, that is, $\Gamma = -d\Theta/dz$ is the *negative* of the vertical temperature gradient, and Γ_a is the *adiabatic lapse rate*, that is, Γ_a is the lapse rate of an atmosphere in adiabatic equilibrium. Invent or find a 'physical' justification of this criterion for stability of an atmosphere (of a planet or star).

[Hint: $P = R\bar{\rho}\Theta$, $R = c_p - c_v$, $\gamma = c_p/c_v$ for a perfect gas, and $P/\bar{\rho}^\gamma$ is independent of height for adiabatic equilibrium.]

3.8 *Kelvin–Helmholtz instability with a three-dimensional basic flow.* Given a basic flow with velocity, density and pressure

$$\mathbf{U} = \begin{cases} U_2\mathbf{i} + V_2\mathbf{j} \\ U_1\mathbf{i} + V_1\mathbf{j}, \end{cases} \quad \bar{\rho} = \begin{cases} \rho_2 \\ \rho_1, \end{cases} \quad P = \begin{cases} p_0 - g\rho_2 z \\ p_0 - g\rho_1 z \end{cases} \begin{matrix} \text{for } z > 0 \\ \text{for } z < 0, \end{matrix}$$

show that normal modes proportional to $e^{i(kx+ly)+st}$ have dispersion relation

$$s = -\frac{i[k(\rho_1 U_1 + \rho_2 U_2) + l(\rho_1 V_1 + \rho_2 V_2)]}{\rho_1 + \rho_2}$$

$$\pm \left\{ \frac{\rho_1 \rho_2}{(\rho_1 + \rho_2)^2} [k(U_2 - U_1) + l(V_2 - V_1)]^2 \right.$$

$$\left. -\frac{(\rho_1 - \rho_2)g(k^2 + l^2)^{1/2}}{\rho_1 + \rho_2} \right\}.$$

Deduce that the flow is stable to the mode if

$$(\rho_1 + \rho_2)(\rho_1 - \rho_2)g > \rho_1 \rho_2 (\Delta U)^2 (k^2 + l^2)^{1/2},$$

where $\Delta U = [k(U_2 - U_1) + l(V_2 - V_1)]/(k^2 + l^2)^{1/2}$ is the basic velocity difference in the direction of the wavenumber vector. [After S. D. Mobbs (private communication).]

3.9 *Saffman–Taylor instability, or Rayleigh–Taylor instability in a porous medium.* Consider the instability of a basic flow in which two incompressible viscous fluids move with a horizontal interface and uniform vertical velocity in a uniform porous medium. You are given that the motion of a fluid in a porous medium is governed by *Darcy's law*, namely that $\mathbf{u} = \nabla \phi$, where $\phi = -k(p + g\rho z)/\mu$, ρ is the density of the fluid, μ its dynamic viscosity, and k is a constant called the *permeability* of the medium to the fluid.

Let the lower fluid have density ρ_1 and viscosity μ_1, and the upper fluid ρ_2 and μ_2; let the medium have permeability k_1 to the lower fluid and k_2 to the upper; and let the basic velocity of the fluids be $W\mathbf{k}$. Then show that the flow is stable if and only if

$$\left(\frac{\mu_1}{k_1} - \frac{\mu_2}{k_2} \right) W + g(\rho_1 - \rho_2) \geq 0.$$

[After Drazin & Reid (1981, Problem 1.11); Saffman & Taylor (1958). Hint: Take the equation of the *mean* interface to be $z = 0$ instantaneously and of the disturbed interface to be $z = \zeta$. Deduce that $\Delta \phi = 0$ where $z \neq \zeta$. Use the continuity of normal velocity and pressure, and linearize the problem. Taking normal modes with $\zeta \propto \exp(st + i\alpha x)$ and so forth, show that

$$\frac{s}{\alpha} \left(\frac{\mu_1}{k_1} + \frac{\mu_2}{k_2} \right) = -g(\rho_1 - \rho_2) - \left(\frac{\mu_1}{k_1} - \frac{\mu_2}{k_2} \right) W.]$$

4

Capillary Instability of a Jet

Another classic problem that does not demand a lot of mathematics is posed and solved in this chapter. Again, it is an idealized model of an important physical phenomenon. It is the problem of the break-up of a round jet of a liquid due to surface tension.

4.1 Rayleigh's Theory of Capillary Instability of a Liquid Jet

Let us take another classic example, first solved by Rayleigh in 1879, to illustrate further the ideas and methods of Chapter 2. Look at the picture of Figure 4.1 to see how a jet of water breaks up into droplets. It looks like an instability. We shall model a liquid jet in air (for example, the water falling from a half-open tap) by a cylinder of an incompressible inviscid liquid, moving with uniform velocity parallel to the axis of the cylinder. Suppose further that there is surface tension, with an ambient fluid of zero density outside the cylinder. Now the most difficult part, namely the modelling of a phenomenon by a tractable mathematical problem, of our task is already done!

It is convenient to use cylindrical polar coordinates (x, r, θ). Then the dimensional equations governing the disturbed jet are as follows. Euler's equations of motion of an inviscid fluid are

$$\rho \left(\frac{\partial \mathbf{u}}{\partial t} + \mathbf{u} \cdot \nabla \mathbf{u} \right) = -\nabla p. \tag{4.1}$$

The equation of continuity for an incompressible fluid, as usual, gives

$$\nabla \cdot \mathbf{u} = 0. \tag{4.2}$$

Laplace's result (see Exercise 3.1) that the drop in pressure across the surface of the liquid jet is the product of the surface tension γ and the sum of the principal curvatures gives

$$p = p_\infty + \gamma \nabla \cdot \mathbf{n} \quad \text{at } r = \zeta, \tag{4.3}$$

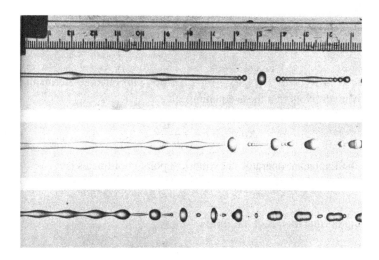

Figure 4.1   The break-up of a jet of water, as different wavelengths are excited. (From Van Dyke, 1982, Fig. 122.)

where $r = \zeta(x, \theta, t)$ is the equation of the surface (when disturbed), p_∞ is the ambient pressure, and the unit outward normal vector from the surface is

$$\mathbf{n} = \frac{(-\partial\zeta/\partial x, 1, -\partial\zeta/r\partial\theta)}{[(\partial\zeta/\partial x)^2 + 1 + (\partial\zeta/r\partial\theta)^2]^{1/2}}. \tag{4.4}$$

We also need the exact kinematic condition that each particle on the surface remains there:

$$u_r = \frac{D\zeta}{Dt} \quad \text{at } r = \zeta. \tag{4.5}$$

To describe the basic state, first make, without loss of generality, a Galilean transformation that reduces the basic flows to rest. Then we may take

$$\mathbf{U} = \mathbf{0}, \qquad P = p_\infty + \gamma/a \quad \text{for } 0 \le r \le a, \tag{4.6}$$

where a is the radius of the jet, γ is its surface tension and p_∞ is the ambient pressure, because $\nabla \cdot \mathbf{n} = \partial n_x/\partial x + \partial(r n_r)/r\partial r + \partial n_\theta/r\partial\theta, = 1/r$ when $\mathbf{n} = (0, 1, 0)$.

To find the stability characteristics, we linearize the problem for small perturbations $\mathbf{u}' = \mathbf{u}$, $p' = p - P$, $\zeta' = \zeta - a$. Thus we find

$$\rho\frac{\partial\mathbf{u}'}{\partial t} = -\nabla p', \tag{4.7}$$

$$\nabla \cdot \mathbf{u}' = 0, \tag{4.8}$$

$$u_r' = \frac{\partial \zeta'}{\partial t}, \quad p' = -\gamma \left(\frac{\zeta'}{a^2} + \frac{\partial^2 \zeta'}{\partial x^2} + \frac{1}{a^2} \frac{\partial^2 \zeta'}{\partial \theta^2} \right) \quad \text{at } r = a, \tag{4.9}$$

because $\nabla \cdot \mathbf{n} = 1/r - \partial^2 \zeta'/\partial x^2 - \partial^2 \zeta'/r^2 \partial \theta^2$ plus quadratic terms in ζ' and its derivatives. Note that these equations give

$$\Delta p' = -\rho \frac{\partial (\nabla \cdot \mathbf{u}')}{\partial t} = 0, \tag{4.10}$$

where the Laplacian operator in cylindrical polar coordinates is

$$\Delta = \partial^2/\partial x^2 + \partial^2/\partial r^2 + \partial/r\partial r + \partial^2/r^2 \partial \theta^2.$$

Taking normal modes of the form

$$(\mathbf{u}', p', \zeta') = \left(\hat{\mathbf{u}}(r), \hat{p}(r), \hat{\zeta} \right) e^{st + \mathrm{i}(kx + n\theta)},$$

we see at once from equation (4.10) that

$$\frac{\mathrm{d}^2 \hat{p}}{\mathrm{d}r^2} + \frac{1}{r} \frac{\mathrm{d}\hat{p}}{\mathrm{d}r} - \left(k^2 + \frac{n^2}{r^2} \right) \hat{p} = 0. \tag{4.11}$$

This is the modified Bessel equation of order n for the functions $I_n(kr)$, $K_n(kr)$; we may take $n \geq 0$ without loss of generality. We deduce that

$$\hat{p}(r) = A I_n(kr) \tag{4.12}$$

for some constant A, in order that $\hat{p}(r)$ is bounded as $r \to 0$. Equation (4.7) now gives

$$\hat{\mathbf{u}} = -A(\rho s)^{-1} \left(\mathrm{i}k I_n(kr), k I_n'(kr), \mathrm{i}n r^{-1} I_n(kr) \right). \tag{4.13}$$

Finally, the linearized boundary conditions give

$$A I_n(\alpha) = -\gamma \left(1 - \alpha^2 - n^2 \right) \hat{\zeta}/a^2, \quad -A(a\rho s)^{-1} \alpha I_n'(\alpha) = s \hat{\zeta}, \tag{4.14}$$

where $\alpha = ak$. Eliminating A, we deduce the eigenvalue relation,

$$s^2 = \frac{\gamma}{a^3 \rho} \frac{\alpha I_n'(\alpha)}{I_n(\alpha)} \left(1 - \alpha^2 - n^2 \right). \tag{4.15}$$

The properties of the Bessel functions give $\alpha I_n'(\alpha)/I_n(\alpha) > 0$ for all $\alpha \neq 0$. Therefore $s^2 < 0$ for all α if $n \neq 0$, but $s^2 > 0$ for $-1 < \alpha < 1$ and $s^2 \leq 0$ for $\alpha \geq 1$ or $\alpha \leq -1$ if $n = 0$. Therefore the jet is stable $\mathrm{Re}(s) = 0$ to all non-axisymmetric modes but unstable to axisymmetric modes whose wavelength

$\lambda = 2\pi/k$ is greater than the circumference $2\pi a$ of the jet. If we define k_m as the value of k at which Re (s) is greatest, then a little calculation gives $k_m \approx 0.7/a$.

What is the nonlinear development of this instability? There is no bifurcation (in *this* model because jets of *all* radii are unstable: the instability just grows faster for thinner jets). There being no dimensionless combination of the physical quantities a, γ, ρ specifying the model, no critical value of a parameter can mark the margin of stability. We expect a small initial disturbance of the jet to excite modes of all wavenumbers k and n. Only those modes with $n = 0$ and $0 < ak < 1$ grow, at exponential rates, so after a little time these modes will leave behind the others, with the most rapidly growing mode ahead of them all. The exponential growth given by the linear theory above cannot last long because, as the disturbance grows, nonlinearity becomes significant. Without numerical calculation of the strongly nonlinear development of a disturbance, we can only look at experimental results or venture to predict that the liquid jet will break up with a wavelength of about $2\pi/k_m \approx 9a$. Go to your kitchen, turn on the water tap, and check this for yourself! You also might look at the photographs of a careful laboratory experiment (Figure 4.1). Even better are the motion pictures of Trefethen (FL1965, edited from F1965), taken by high-speed photography with carefully controlled lighting. Figure 4.2 gives some quantitative verification of Rayleigh's theory.

This problem has been selected here as a simple illustration of the methods of the linear theory of hydrodynamic stability that can be directly related to

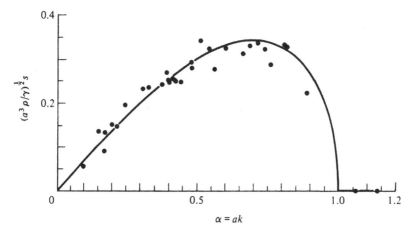

Figure 4.2   Graph of the dimensionless growth rate $(a^3\rho/\gamma)^{1/2}s$ against the dimensionless wavenumber α for axisymmetric capillary modes $(n = 0)$. The curve denotes Rayleigh's theoretical results and the points the experimental results measured by Donnelly & Glaberson (1966).

physical observation. However, it is fundamental to important problems of atomization, with drop and spray formation. These wider aspects and more recent work on the break-up of jets have been reviewed by Lin & Reitz (1998).

Exercises

4.1 *The capillary instability of a jet of gas.* For a gas jet in liquid with the liquid at rest in the region $r > a$ *outside* the gas, show that the eigenvalue relation (4.15) is replaced by

$$s^2 = -\frac{\gamma}{a^3 \rho} \frac{\alpha K_n'(\alpha)}{K_n(\alpha)} \left(1 - \alpha^2 - n^2\right).$$

Hence show that there is instability only for the varicose mode ($n = 0$) with $-1 < \alpha < 1$, and that the dimensionless relative growth rate $(a^3 \rho / \gamma)^{1/2} s$ attains its maximum value 0.820 for $\alpha = 0.484$. [After Drazin & Reid (1981, Problem 1.8).]

4.2 *The role of vorticity in capillary instability.* Equation (4.7) implies that the vorticity $\boldsymbol{\omega}' = \nabla \times \mathbf{u}'$ is independent of time. Deduce that

$$\boldsymbol{\omega}'(\mathbf{x}, t) = \boldsymbol{\omega}'(\mathbf{x}, 0).$$

Discuss the significance of this in the linearized initial-value problem of capillary instability, noting that the irrotational and rotational components of a perturbation may be superposed. [After Drazin & Reid (1981, Problem 1.9).]

4.3 *The capillary stability of a plane jet.* An incompresible inviscid liquid with density ρ and surface tension γ is at rest between the free plane surfaces $z = \pm a$. Show that the stability of the liquid is governed by the equation

$$\Delta p' = 0 \quad \text{for} -\infty < x, y < \infty, -a < z < a,$$

and the boundary conditions

$$\rho \frac{\partial^2 p'}{\partial t^2} = \pm \gamma \left(\frac{\partial^2}{\partial x^2} + \frac{\partial^2}{\partial y^2} \right) \frac{\partial p'}{\partial z} \quad \text{at } z = \pm a.$$

Taking normal modes with $p' \propto \exp[st + i(kx + ly)]$, show that

$$\frac{a^3 \rho s^2}{\gamma} = -\tilde{\alpha}^3 \tanh \tilde{\alpha} \quad \text{or} \quad -\tilde{\alpha}^3 \coth \tilde{\alpha},$$

where $\tilde{\alpha} = a(k^2 + l^2)^{1/2}$. Deduce that this plane jet is stable, unlike the round jet, which is unstable to axisymmetric disturbances.

[After Drazin & Reid (1981, Problem 1.10). Hint: $\rho \partial \mathbf{u}'/\partial t = -\nabla p'$, $\nabla \cdot \mathbf{u}' = 0$; and $\mathbf{n} = \pm(-\partial \zeta'/\partial x, -\partial \zeta'/\partial y, 1)$, $p' = -\gamma \nabla \cdot \mathbf{n}$, $w' = -\partial \zeta'/\partial t$, at $z = \pm a$. Note that $s^2 \sim -\gamma \tilde{\alpha}^3/a^3 \rho$ as $\tilde{\alpha} \to \infty$ for short capillary waves at a free surface, in common with the round jet.]

5

Development of Instabilities in Time and Space

And wisdom and knowledge shall be the stability of thy times. . . .
Isaiah xxxiii 6

More advanced properties of instabilities will be described in this chapter. The development of normal modes in space as well as time, the effect of weak nonlinearity and the energy budget will be explained.

5.1 *The Development of Perturbations in Space and Time

For partial differential systems, such as those describing fluid motions, it is valuable to analyse the nature of stability in more detail.

First, note that if a flow is bounded (and, of course, in practice *all* flows are bounded), then there is in general a countable infinity of normal modes, but that if the flow is unbounded then there is an uncountable infinity of normal modes; for the Poiseuille pipe flow of Example 2.11, which is unbounded in the x-direction, there is a continuum of modes with a continuous wavenumber k as well as discrete wavenumbers for θ- and r-variations, but for flow in a cube there would be three discrete wavenumbers to specify each normal mode. So for an unbounded flow the most unstable mode can be no more than first among equals, but for a bounded flow the growth rate of the most unstable mode will in general be substantially greater than that of the second most unstable mode. For bounded flows of large aspect ratio (or large Reynolds number), the most unstable modes are usually close together and so approximate a continuum.

It is also helpful to distinguish between what are often called closed and open flows. No rigorous definitions have been universally accepted, but a flow in which all the fluid particles passing through each point return there is said to be *closed*. A flow in which fluid particles enter or leave the domain of flow through the boundaries or at infinity is said to be *open*. For example, the state of rest of a fluid in an impermeable box and Couette flow between two long rotating coaxial cylinders are closed flows. Also Poiseuille flow along an infinitely long circular pipe is an open flow; so are channel flows, Blasius's boundary layer on a flat plate, a jet, a wake, a vortex street and a free shear (that is, a mixing) layer. An open flow often has a natural origin like the leading edge of the plate to which the boundary layer is attached, the nozzle of the jet, or the trailing edge of a splitter plate which produces the free shear layer; in this event the open flow is not Galilean invariant.

An unbounded flow admits physically different types of instability. It is *absolutely unstable* if a sufficiently small perturbation grows above a given threshold at a fixed point of the flow. We may express this more mathematically by defining the flow to be absolutely unstable if there exists a small initial perturbation satisfying the linearized problem such that

$$|\mathbf{u}'(\mathbf{x}, t)| \to \infty \quad \text{as } t \to \infty$$

for all fixed $\mathbf{x}$. The flow is *convectively unstable* if similarly a sufficiently small perturbation grows above a given threshold at no fixed point of the flow but does grow at a moving point; in this case the perturbation may grow exponentially while it propagates downstream (and out of the laboratory waste pipe). This requires that

$$|\mathbf{u}'(\mathbf{x}, t)| \to 0 \quad \text{as } t \to \infty$$

for all fixed $\mathbf{x}$ but there exists $\mathbf{V}$ such that

$$|\mathbf{u}'(\mathbf{x} + \mathbf{V}t, t)| \to \infty \quad \text{as } t \to \infty.$$

Note that this distinction is not Galilean invariant. Convective and absolute instabilities are illustrated in Figure 5.1.

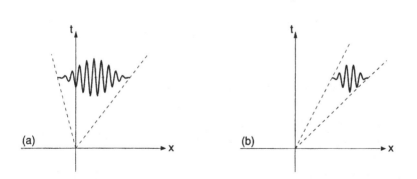

Figure 5.1 Symbolic sketch of the development of (a) absolute and (b) convective instability in the (x, t)-plane. (After Schmid & Henningson, 2001, Fig. 7.6; reproduced by permission of Springer-Verlag GmbH & Co. KG.)

Example 5.1: Dispersion. To introduce dispersion and a little notation, take the linear equation

$$\frac{\partial u'}{\partial t} + W\frac{\partial u'}{\partial x} = \frac{\partial^2 u'}{\partial x^2} + (R - R_c)u', \qquad (5.1)$$

for constant W, merely as a toy problem. We may envisage it as the linearized equation for stability of a basic 'flow' U. This admits normal modes of the form

$$u'(x, t) = \mathrm{Re}\left[Ae^{i(kx-\omega t)}\right], \qquad (5.2)$$

where the *dispersion relation* of the frequency to the wavenumber is

$$\mathcal{D}(k, \omega; R) = 0, \qquad (5.3)$$

and $\mathcal{D}(k, \omega; R) = -i\omega + iWk + k^2 - (R - R_c)$ for equation (5.1). It follows that $\omega = f(k)$, where $f(k) = Wk - ik^2 + i(R - R_c)$ so that $W = f'(0)$, $R - R_c = -if(0)$. Also equation (5.1) may be symbolically written as $\mathcal{D}(-i\partial/\partial x, i\partial/\partial t; R)u' = 0$.

The 'flow' is asymptotically stable if $R < R_c$ because all modes decay exponentially. The 'flow' is convectively unstable if $R > R_c$ and $W \neq 0$ because the modes which grow exponentially also propagate. The 'flow' is absolutely unstable if $R > R_c$ and $W = 0$. The group velocity of waves (5.2) is $c_g = \partial f/\partial k = W - 2ik$. $\square$

Another important idea, coming from plasma physics, is useful in considering the perturbations generated by vibrating ribbons or small loudspeakers, which are inserted by experimentalists into flows in channel or boundary layers to create a small source of oscillations at a fixed frequency. If the flow is absolutely unstable, then insertion of the source will at once generate perturbations which grow linearly in the locality of the source, where one might expect nonlinearity and perhaps turbulence to occur soon. If the flow is stable or convectively unstable, then the perturbation may have the same frequency as the source in the locality of the source, where they decay or grow in *space*. These are called *spatial modes*. If the source has frequency ω, then the perturbation may grow or decay in space like e^{ikx}, where k is a root of $\mathcal{D}(k, \omega; R) = 0$.

It may be anticipated that in general there will be an infinite spectrum of such complex roots k, and that, of the corresponding normal modes, the fastest growing will dominate the perturbation soon after the source is turned on, such that the perturbation lies downstream or upstream of the source according to the sign of the group velocity of the mode.

To substantiate these ideas simply, take another toy problem in one dimension, say a basic 'flow' U with perturbation $u'(x, t)$ and some dispersion function $\mathcal{D}(k, \omega)$. Then we can reconstruct the linearized equation

$$\mathcal{D}(-i\partial/\partial x, i\partial/\partial t)u'(x, t) = 0. \tag{5.4}$$

To examine the development of the perturbations due to various sources, consider the Green's function G due to an instantaneous localized unit source at $t = 0$, $x = 0$. It satisfies the inhomogeneous equation

$$\mathcal{D}(-i\partial/\partial x, i\partial/\partial t)G(x, t) = \delta(t)\delta(x) \tag{5.5}$$

and vanishes for all $t < 0$ and as $x \to \pm\infty$ for fixed $t > 0$. It is instructive to note the behaviour of G as $t \to \infty$ for fixed $x/t, = V$, say. If the basic flow is asymptotically stable, then

$$G(x, t) \to 0 \quad \text{as } t \to \infty \; \forall V. \tag{5.6}$$

If the basic flow is exponentially unstable, then

$$G(x, t) \to \infty \quad \text{as } t \to \infty \tag{5.7}$$

for at least one value of V. If, moreover, the basic flow is convectively unstable, then

$$G(x, t) \to 0 \quad \text{as } t \to \infty \text{ for } V = 0; \tag{5.8}$$

and if the basic flow is absolutely unstable, then

$$G(x, t) \to \infty \quad \text{as } t \to \infty \text{ for } V = 0. \tag{5.9}$$

Such consideration of a linearized initial-value problem and study of the solution after a long time was initiated by Landau (1946) and comprehensively developed by Briggs (1964) for plasma physics. Similar ideas have subsequently been applied to linearized problems of hydrodynamic stability, notably by Huerre & Monkewitz (1990). Of course, in both subjects the behaviour of perturbations after a long time is physically important. There are many technical complications according to the nature of the dispersion function $\mathcal{D}$, so only the fundamentals are summarized here. By use of Laplace and Fourier transforms, the Green's function can be expressed as

$$G(x, t) = \frac{1}{(2\pi)^2} \int_F \int_L \frac{e^{i(kx-\omega t)}}{\mathcal{D}(k, \omega)} \, d\omega \, dk, \tag{5.10}$$

where F is an appropriate contour in the complex k-plane to invert the Fourier transform in k and L one in the ω-plane to invert the Laplace transform. The location and manipulation of the contours F and L is a subtle technical issue, dependent on the nature and location of the singularities of the integral and so on the zeros of the dispersion function $\mathcal{D}$. The behaviour of $G(x, t)$ as $t \to \infty$ can be found by the method of steepest descent. The saddle points are found to occur where

$$\frac{x}{t} = \frac{d\omega}{dk}(k),$$

the group velocity, where $\mathcal{D}(k, \omega(k)) = 0$ defines a branch of the function $\omega(k)$. In many problems of interest the complex root k_0 of

$$\frac{d\omega}{dk}(k_0) = 0$$

can be defined, so that it can be shown that if $\mathrm{Im}\,(\omega(k_0)) > 0$, then the flow is absolutely unstable, and if $\mathrm{Im}\,(\omega(k_0)) < 0$, then the flow is convectively unstable.

Similarly, the linearized perturbation due to a localized unit source at $x = 0$ which has real frequency ω and is turned on at $t = 0$ can be expressed as an integral by use of Fourier and Laplace transforms, and the integral evaluated asymptotically as $t \to \infty$ for all fixed x. This gives spatial modes of frequency ω downstream or upstream of the source according to the sign of the group velocity if the flow is not absolutely unstable.

Initial-value linearized problems of hydrodynamic stability in three dimensions have much more structure than these one-dimensional toy problems, but similar principles apply to both sets of problems. It is often found for flows of a viscous fluid that all modes are stable for $R < R_c$, some are convectively unstable for $R_c < R < R_a$, but some are absolutely unstable only for $R > R_a$, where R_a is some value of R, the Reynolds number or other governing dimensionless parameter.

A very different approach to the development of perturbations in space and time has been applied to the evolution of wave packets comprised of weakly unstable modes. The approach can be sketched intuitively as follows, again with only one space dimension, so that the independent variables are only x, t. Many of the ideas can be generalized to higher space dimensions with little trouble. There is a wide variety of problems of hydrodynamic stability which are both invariant under reflection in the x-axis and homogeneous in space, so we will assume these properties. Then let the linearized problem admit normal modes of the form

$$u'(x, t) = \mathrm{Re}\big(Ae^{ikx+st}\big), \tag{5.11}$$

for real wavenumbers k, with dispersion relation $s = \sigma + i\tau = f(k, R)$, say, where the problem is governed by some parameter R such as a Reynolds number. The essential idea is to approximate the dispersion function $\mathcal{D}(k, s; R)$ algebraically for $0 < R - R_c \ll 1$ to describe the narrow band of weakly unstable modes, and then reconstruct the approximate partial differential equation satisfied by u'.

Now, it follows from the invariance under reflection that $Ae^{-ikx+st}$ is also a solution of the linearized problem and therefore that $f(-k, R) = f(k, R)$. Let us further assume that there is a most unstable or least stable mode of wavenumber $k_c(R) \neq 0$ and that f is an analytic function of k at k_c; it follows that $\sigma = \text{Re}(f)$ has a simple maximum at $k = k_c(R)$. Therefore

$$\sigma = a - b\left(k^2 - k_c^2\right)^2 + O\left[\left(k^2 - k_c^2\right)^3\right] \quad \text{as } k \to k_c,$$

where $a \sim \gamma(R - R_c), 0 < b \to b_c$ as $R \to R_c$ for $k > 0$. A marginal curve $\sigma(k, R) = 0$ is sketched in Figure 5.2(a), characteristic of many problems of hydrodynamic stability. The graph of σ as a function of k for fixed R close to R_c then has the shape sketched in Figure 5.2(b). Note that the width of the band of unstable waves is of order of magnitude ϵ as $\epsilon \to 0$, where $\epsilon = (R - R_c)^{1/2}$.

Also let $\omega = -\text{Im}[f(k_c, R_c)]$. Then the problem for $0 < R - R_c \ll 1$ admits approximate modes of the form

$$u'(x, t) = \text{Re}\{A(X, T) \exp[i(k_c x - \omega t)]\}, \tag{5.12}$$

where X, T are some 'slow' coordinate and time respectively. It follows that $X = \epsilon^2 x, T = \epsilon^2 t$ and A satisfies an evolution equation whose terms correspond to those in equation (5.3). By making a Galilean transformation whereby the flow is measured in a frame moving with the group velocity $c_g = -[\partial\tau/\partial k]_{R=R_c, k=k_0}$ of the most unstable mode at marginal stability, the term in $\partial A/\partial X$ may be removed. Here we define $k_0 = k_c(R_c)$. It thus follows that modulations of the most unstable mode are governed by the equation

$$\frac{\partial A}{\partial T} = \gamma A + 4\iota k_0^2 b_c \frac{\partial^2 A}{\partial X^2}. \tag{5.13}$$

This partial differential equation governs approximately the modulation in space and time of a growing packet of weakly unstable waves while the packet propagates at its group velocity, but will of course break down as soon as nonlinearity becomes significant.

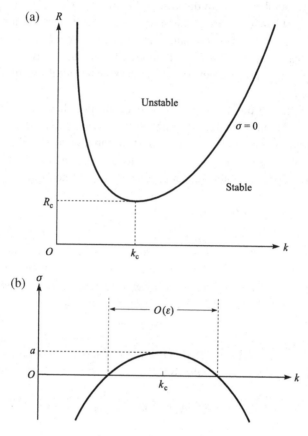

Figure 5.2   (a) Symbolic sketch of a typical marginal curve $\sigma = 0$ in the (k, R)-plane. (b) Sketch of the graph of $\sigma(k, R)$ for fixed R such that $0 \ll \epsilon^2 = R - R_c \ll 1$.

5.2 Weakly Nonlinear Theory

Another important method of the theory of hydrodynamic stability is weakly nonlinear theory, essentially a method of perturbing the linear stability characteristics for disturbances of small amplitude. It also gives the local theory of a bifurcation of flow regime.

We have seen in §1.1 that Reynolds (1883) appreciated the physical importance of nonlinear disturbances of Poiseuille flow in a pipe. The first steps of a nonlinear theory were taken by Stokes (1880) for surface gravity waves in deep water, Bohr (1909) for capillary instability, and Noether (1921) and Heisenberg (1924) for instability of plane parallel flows of a viscous fluid, but the earliest enduring results for instability are those of Landau (1944).

He had a vision of the weakly nonlinear theory which led him to write down what is now called the Landau equation, but he did not derive it directly for the instability of any specific flow. The *steady* form of his equation was first derived from a problem of fluid mechanics independently by Gor'kov (1957) and Malkus & Veronis (1958), who solved the weakly nonlinear problem of Rayleigh–Bénard convection with free–free boundary conditions. Stuart (1960) and Watson (1960b) first derived asymptotically the full Landau equation, for weakly unstable wave disturbances of plane parallel flows.

First it can be seen, in a general way, how a small weakly unstable perturbation may generate weakly nonlinear interactions and thereby quench its own instability. In the linear theory of stability of a steady basic flow, with velocity $\mathbf{U}$ say, the perturbed flow, $\mathbf{u} = \mathbf{U} + \mathbf{u}'$ say, is considered, the equations of motion linearized for a small perturbation, $\mathbf{u}'$, and the perturbation expressed as a superposition of normal modes of the form $\mathbf{u}' = \epsilon[A(t)\hat{\mathbf{u}} + A^*(t)\hat{\mathbf{u}}^*]$, say, for some small parameter ϵ such that the complex amplitude $A(t) = A_0 e^{st}$ is of order one for moderate values of time t.

It is usually a valid approximation to consider the evolution of the most unstable mode and ignore the others when the flow is only just unstable, so that there is either only one unstable mode or only a single narrow band of unstable modes in a wave packet. Now the Navier–Stokes equations are quadratically nonlinear, so that the nonlinear part $\mathbf{u}' \cdot \nabla \mathbf{u}'$ of their nonlinear term $\mathbf{u} \cdot \nabla \mathbf{u}$ will generate terms proportional to $|A|^2$ and A^2, A^{*2} at order ϵ^2. At first these terms will grow exponentially with A. They soon generate a cascade of further nonlinear interactions with one another and the normal mode. At some order, in general quadratic, of ϵ there is a resonant interaction with the normal mode which will change its rate of growth. Then, the major effect of weak nonlinearity will be to have altered the slow exponential rate of growth of the unstable linear mode but not to have changed its spatial character.

It is possible to describe the cascade of nonlinear interactions more fully when the normal mode is a plane wave with $\mathbf{u}' = \epsilon[A(t)e^{i\alpha x}\hat{\mathbf{u}}(y, z) + A^*(t)e^{-i\alpha x}\hat{\mathbf{u}}^*(y, z)]$, say. Then the quadratic terms generated will be proportional to $|A|^2$ and $A^2 e^{2i\alpha x}$, $A^{*2}e^{-2i\alpha x}$. We call the wave mode the *fundamental*, and the contribution to the flow represented by terms in $e^{\pm 2i\alpha x}$ the *first harmonic*. Thus the fundamental both generates the first harmonic, which has half its wavelength and grows at twice its relative rate, and modifies the mean flow (as represented by the term proportional to $|A|^2$ and independent of x). Now the first harmonic and the modification of the mean flow in their turn generate, through the quadratic term in the Navier–Stokes equations, terms proportional both to $|A|^2 A e^{i\alpha x}$, $|A|^2 A^* e^{-i\alpha x}$, and to $A^3 e^{3i\alpha x}$, $A^{*3}e^{-3i\alpha x}$, at order ϵ^3. (The implied translational symmetry of the flow is the cause of the nonlinearity having its leading effect at cubic rather than quadratic order in ϵ.) The latter pair of terms represent the generation of the *second harmonic*, which has a third of the wavelength of the fundamental. The former pair of terms interact resonantly with the fundamental and may enhance or moderate its exponential rate of growth; in particular, this interaction may equilibrate the fundamental so that it becomes a steady nonlinear wave. It is by this mechanism that a weakly unstable normal mode of small magnitude may grow exponentially for a long time and eventually generate weak but important nonlinearity which quenches the growth of the fundamental.

The technical details of the weakly nonlinear theory for hydrodynamic problems are severe, and not very suitable for a first course on the subject, so we shall resort to a simple model problem to illustrate the fundamental ideas of the theory.

Example 5.2: A model problem of a nonlinear diffusion equation. Consider the equation

$$\frac{\partial u}{\partial t} - \sin u = \frac{1}{R}\frac{\partial^2 u}{\partial z^2}, \tag{5.14}$$

together with the boundary conditions

$$u = 0 \quad \text{at } z = 0, \pi, \tag{5.15}$$

after Matkowsky (1970). You may regard this problem as a model of the flow of a fluid with velocity $u(z, t)$ along a channel between the parallel walls $z = 0$ and $z = \pi$, where R is a 'Reynolds' number.

First take as a basic 'flow' the solution $u = U$, where U is the null solution such that $U(z) = 0$ for all z. Next suppose that the basic flow is perturbed such that $u = U + \epsilon u' + o(\epsilon)$, $u' = O(1)$ as $\epsilon \to 0$, and the constant ϵ may be regarded as a small amplitude of the perturbation. Now substitute this form of solution into equation (5.14) and boundary conditions (5.15), and equate coefficients of ϵ, that is, linearize the system, to find

$$\frac{\partial u'}{\partial t} - u' = \frac{1}{R}\frac{\partial^2 u'}{\partial z^2}, \tag{5.16}$$

$$u' = 0 \quad \text{at } z = 0, \pi. \tag{5.17}$$

The solution of this linearized problem may be represented as the Fourier series

$$u'(z, t) = \sum_{n=1}^{\infty} A_n(t) \sin nz. \tag{5.18}$$

Now the integral of the product of $\sin mz$ and equation (5.16) from $z = 0$ to π gives, on integration by parts twice and use of boundary conditions (5.17),

$$\frac{dA_m}{dt} - A_m = -\frac{m^2}{R} A_m. \qquad (5.19)$$

This gives the normal mode with

$$A_m(t) = A_m(0) \exp(s_m t), \qquad s_m = 1 - m^2/R.$$

It follows that the mth mode is stable if $R < m^2$ and the basic 'flow' is stable if all modes are stable, and so if

$$R < R_c = \min_{m \geq 1} m^2 = 1.$$

The 'flow' similarly is unstable to at least one mode if $R > 1$.

These linear results could have been deduced more simply, albeit more naively, by (i) directly taking the normal mode $u'(z, t) = \hat{u}(z)e^{st}$, (ii) finding from equations (5.16), (5.17) that

$$\frac{d^2\hat{u}}{dz^2} + R(1 - s)\hat{u} = 0,$$

$$\hat{u} = 0 \quad \text{at } z = 0, \pi;$$

and (iii) deducing that

$$\hat{u}(z) = \sin mz, \qquad s = 1 - m^2/R,$$

as above.

Next consider the weakly nonlinear development of unstable perturbations when $0 < R - R_c \ll 1$, so that the basic 'flow' is weakly unstable. You may look at the problem heuristically as follows. All modes with $m \geq 2$ are strongly damped when R is close to R_c, so all components of an initial small perturbation, except the first component A_1, will die out rapidly. If $A_1(0)$ is small enough, A will at first grow exponentially as given by the linear theory, with

$$\frac{d(\epsilon A_1)}{dt} = s_1 \epsilon A_1$$

approximately. However, after a long time of order $1/s_1 \approx 1/(R - R_c)$, the amplitude $A_1(t)$ will no longer be small, and nonlinearity will begin to affect its evolution. We may anticipate that terms nonlinear in A_1 would have to be added to the linear amplitude equation above in order to describe the evolution of A_1 for such large times. Note that the original nonlinear problem (5.14),

(5.15) is invariant under translations of the coordinate z by π, and that the translation changes the sign of $\sin z$, which is equivalent to a change of sign of A_1. So we anticipate that the amplitude equation for A_1 is invariant under change of sign of A_1, and that therefore the nonlinear term of lowest power of A_1 will be cubic, not quadratic, and so that for weakly nonlinear disturbances the amplitude equation is a Landau equation

$$\frac{d(\epsilon A_1)}{dt} = s_1 \epsilon A_1 - l \epsilon^3 A_1^3,$$

for some Landau constant l. To describe the evolution of A_1 over long times, it is necessary that these terms are of the same order of magnitude when ϵ is small, that is, the *distinguished limit* is taken as $\epsilon \to 0$. Therefore we require that $s_1 = O(\epsilon^2)$, and therefore $R - R_c = O(\epsilon^2)$ as $\epsilon \to 0$.

The intuitive approach of the last paragraph suggests that we formally define a *slow time* as

$$T = \epsilon^2 t$$

and expand

$$u(z, t, \epsilon) = U(z) + \epsilon u_1(z, T) + \epsilon^2 u_2(z, T) + \cdots \qquad (5.20)$$

as $\epsilon \to 0$ for fixed z, T, where we define $\epsilon = [(R - R_c)/R_2]^{1/2}$ for R_2 independent of R. We may fix positive ϵ in terms of R by taking $R_2 = \pm 1$ according to whether R is greater or less than R_c. This expansion is motivated by the last paragraph, but its justification will come later when its self-consistency is verified (it is too much to ask for an analytic proof of the convergence of such an asymptotic solution of a problem as intricate even as this).

So next we will use expansion (5.20) to solve the problem and check that it is indeed self-consistent. First re-write the problem, without approximation, in terms of T so that the smaller terms are on the right-hand side of the equation,

$$\frac{\partial^2 u}{\partial z^2} + R_c u = R \frac{\partial u}{\partial t} + R_c(u - \sin u) - (R - R_c) \sin u,$$

$$= \epsilon^2 R \frac{\partial u}{\partial T} + R_c(u - \sin u) - \epsilon^2 R_2 \sin u, \qquad (5.21)$$

together with the boundary conditions (5.15). It only remains to equate coefficients of successive powers of ϵ, and solve the resultant succession of *linear* problems that arise, finding $U, R_c, u_1, u_2, \ldots$ in turn. Putting $\epsilon = 0$, which is equivalent to equating coefficients of ϵ^0, gives nothing except confirmation that the null function U is indeed a solution of the problem.

Equating coefficients of ϵ, we find that

$$\frac{\partial^2 u_1}{\partial z^2} + R_c u_1 = 0$$

and u_1 satisfies conditions (5.15). This gives

$$R_c = 1, \qquad u_1(z, T) = A(T) \sin z$$

for the least stable mode, as before in the linear problem. The function A is arbitrary at this stage, but will be determined later.

Equating coefficients of ϵ^2, we find that

$$\frac{\partial^2 u_2}{\partial z^2} + u_2 = 0,$$

and u_2 satisfies conditions (5.15). Therefore $u_2(z) = 0$ for all z. (We could take u_2 equal to an arbitrary multiple of u_1, but we may normalize without loss of generality so that this multiple is zero. This will not affect the ultimate solution, but will affect the way we find it.)

Equating coefficients of ϵ^3, we find that

$$\frac{\partial^2 u_3}{\partial z^2} + u_3 = R_c \frac{\partial u_1}{\partial T} + \frac{1}{6} R_c u_1^3 - R_2 u_1$$

$$= \frac{dA}{dT} \sin z + \frac{1}{6}(A \sin z)^3 - R_2 A \sin z$$

$$= \left(\frac{dA}{dT} - R_2 A + \frac{1}{8} A^3 \right) \sin z - \frac{1}{24} A^3 \sin 3z,$$

and u_3 satisfies conditions (5.15). By the Fredholm alternative (see Exercise 5.8), this linear inhomogeneous two-point boundary-value problem has either (i) a unique solution if the associated homogeneous problem has no solution or (ii) no solution or an infinity of solutions if the homogeneous problem has a solution. Therefore it cannot have a unique solution, because we have shown that u_1 is a solution of the homogeneous problem. So a *solvability condition* must be satisfied in order that u_3 exists and the expansion in power series may be valid. The solvability condition is found as follows. The boundary conditions

(5.15) for u_3 give, whatever the unknown function u_3 is,

$$
\begin{aligned}
0 &= \left[\sin z \frac{\partial u_3}{\partial z} - u_3 \frac{\mathrm{d}\sin z}{\mathrm{d}z} \right]_0^\pi \\
&= \int_0^\pi \left(\sin z \frac{\partial^2 u_3}{\partial z^2} - u_3 \frac{\mathrm{d}^2 \sin z}{\mathrm{d}z^2} \right) \mathrm{d}z \\
&= \int_0^\pi \sin z \left(\frac{\partial^2 u_3}{\partial z^2} + u_3 \right) \mathrm{d}z \\
&= \int_0^\pi \sin z \left(\frac{\mathrm{d}A}{\mathrm{d}T} \sin z + \frac{1}{6} A^3 \sin^3 z - R_2 A \sin z \right) \mathrm{d}z \\
&= \frac{1}{2}\pi \left(\frac{\mathrm{d}A}{\mathrm{d}T} + \frac{1}{8} A^3 - R_2 A \right).
\end{aligned}
$$

Therefore

$$
\frac{\mathrm{d}A}{\mathrm{d}T} = R_2 A - \frac{1}{8} A^3. \tag{5.22}
$$

This is the Landau equation, in which $R_2 = \pm 1$ according to whether the 'Reynolds number' R is supercritical or subcritical respectively. You can see that changing the value of R_2 does not change the solution, but merely the means of its expression, by changing the scale of ϵ or making ϵ complex, because the Landau equation above is essentially

$$
\epsilon^3 \frac{\mathrm{d}A}{\mathrm{d}T} = (R - R_\mathrm{c})\epsilon A - \frac{1}{8}\epsilon^3 A^3.
$$

It is possible to go on to find an infinity of solutions u_3 and render u_3 unique by use of a normalization condition, find a solvability condition for the existence of u_4, then u_4 itself, and so forth; but, now the leading asymptotic behaviour of u for small $R - R_\mathrm{c}$ through the Landau equation has been found, it is time to stop. You can see the pattern of the iteration for yourself, and recognize the self-consistency of the procedure.

The leading behaviour of the solution (5.20) as $R \to R_\mathrm{c}$ is that the spatial character of the instability is determined by the eigenfunction (namely $\sin z$) of the most unstable mode of the *linear* problem and the temporal character is determined by the solutions A of the nonlinear amplitude equation (5.22).

As a postscript to this example, reconsider how the solvability condition was found above. What is the reason for starting with the identity $0 = [\sin z \partial u_3/\partial z -u_3 \mathrm{d}\sin z/\mathrm{d}z]_0^\pi$? How could you have thought of this for yourself? If you do not understand the motivation of the choice of the identity, try solving Exercise 5.7. $\square$

Recall that in §5.1 we showed how a weakly unstable wave packet is often described by the linear amplitude equation (5.13). Making the same assumptions about the dispersion relation for a linearized problem of hydrodynamic stability, consider the weakly nonlinear problem of weakly unstable modes, so that $|A|$ is small as well as $R - R_c$. Again take

$$u(x, t) = \text{Re}\,[A(X, T)\exp(i\alpha_c x - i\omega t)], \qquad (5.23)$$

where $X = \epsilon^2 x, T = \epsilon^2 t$. Then A satisfies a nonlinear evolution equation whose *linear* terms are the same as in equation (5.13). Now the assumption of spatial homogeneity implies that the problem is also invariant under translations $x \mapsto x + l$ for all l, for which $A \mapsto A \exp(i\alpha_c l)$. Therefore the leading *weakly nonlinear* terms in the amplitude evolution equation will be cubic and involve only $|A|^2 A$ in order that the equation is invariant under reflection and all translations in x. Also Example 5.2 shows that A is of order ϵ. It thus follows that weakly nonlinear modulations of the most unstable mode are governed by the *Ginzburg–Landau equation*, namely

$$\frac{\partial A}{\partial T} = kA - l|A|^2 A + 4i\alpha_0^2 b_c \frac{\partial^2 A}{\partial X^2}, \qquad (5.24)$$

for some complex number l. This equation is applicable to perturbations of many types of flow (and, indeed, to many nonlinear phenomena which do not concern fluids), although the calculations to evaluate the constants are often lengthy and complicated.

Forms of amplitude evolution equation other than the Landau equation (2.9) may arise in other problems in the weakly nonlinear approximation.

Algebraic terms other than those on the right-hand side of (2.9) may arise, as we have seen in §2.2; the ordinary differential system governing the evolution may be of higher order than the first, with more than one amplitude; the system may be integro-differential.

Further, for disturbances of open flows there is in general weakly nonlinear modulation of the amplitude in space as well as time, as is shown in §5.1 for the linearized problem. Craik (1985) treated these issues at length.

However, by looking at the symmetries of a problem of weakly nonlinear hydrodynamic instability, it may be possible to pass over the long and detailed calculations necessary to find the coefficients of the amplitude equation numerically, and determine the *form* of the amplitude equation with relative ease. Such a form, for example, that of the Landau equation, is called a *normal form*. The qualitative properties of weakly nonlinear disturbances can then be found from the normal form, just as we found them from the Landau equation in §2.2.

5.3 The Equation of the Perturbation Energy

The important physical aspects of the energy budget of the disturbance of a general basic flow are examined in this section. Changing to tensor notation, write

$$\frac{\partial((1/2)\mathbf{u}'^2)}{\partial t} = u_i' \frac{\partial u_i'}{\partial t}$$

$$= -u_i' U_j \frac{\partial u_i'}{\partial x_j} - u_i' u_j' \frac{\partial U_i}{\partial x_j} - u_i' u_j' \frac{\partial u_i'}{\partial x_j} - u_i' \frac{\partial p'}{\partial x_i} + R^{-1} u_i' \frac{\partial^2 u_i'}{\partial x_j^2},$$

on using the *exact* dimensionless equation (2.15) of motion for $\mathbf{u}'$, p'. Therefore

$$\frac{\partial((1/2)\mathbf{u}'^2)}{\partial t} = -\frac{\partial}{\partial x_j}\left(\frac{1}{2}u_i'^2 U_j\right) + D_{ij}(-u_i' u_j') - \frac{\partial}{\partial x_j}\left(\frac{1}{2}u_i'^2 u_j'\right)$$

$$\quad\text{(A)}\qquad\qquad\text{(B)}\qquad\qquad\text{(C)}\qquad\qquad\text{(D)}$$

$$-\frac{\partial(p' u_i')}{\partial x_i} + R^{-1}\left[\frac{\partial}{\partial x_j}\left(u_i' \frac{\partial u_i'}{\partial x_j}\right) - \left(\frac{\partial u_i'}{\partial x_j}\right)^2\right]; \quad (5.25)$$

$$\qquad\qquad\text{(E)}\qquad\qquad\qquad\qquad\text{(F)}$$

because $\nabla \cdot \mathbf{u}' = 0$ and $\nabla \cdot \mathbf{U} = 0$, where the rate-of-strain tensor of the basic flow is

$$D_{ij} = \frac{1}{2}\left(\frac{\partial U_i}{\partial x_j} + \frac{\partial U_j}{\partial x_i}\right). \quad (5.26)$$

Equation (5.25) gives the rate-of-change of the energy of the perturbation. Each of the six terms may be interpreted physically as follows: (A) represents the rate of increase of the kinetic energy density (remember that dimensionless variables with $\rho = 1$ are being used); (B) represents the convection of the perturbation kinetic energy by the basic flow; (C) represents the energy transferred to the perturbation from the basic flow; (D) represents the convection of perturbation kinetic energy by the perturbation velocity; (E) represents the rate of working of the perturbation pressure on the perturbation; and (F) represents the viscous dissipation of energy.

This can be seen more clearly by integrating the energy equation over the domain of flow. First define the total kinetic energy of the perturbation as

$$K = \int_{\mathcal{V}} \frac{1}{2} u_i'^2 \, d\mathbf{x}. \tag{5.27}$$

Now

$$\frac{dK}{dt} = \int_{\mathcal{V}} u_i' \frac{\partial u_i'}{\partial t} \, d\mathbf{x},$$

at least if $\mathcal{V}$ is fixed for all t,

$$= -\int_{\mathcal{V}} \left[D_{ij} u_i' u_j' + R^{-1} \left(\frac{\partial u_i'}{\partial x_j} \right)^2 \right] d\mathbf{x}, \tag{5.28}$$

on use of equation (5.25), Gauss's divergence theorem and the boundary conditions. This is the famous *Reynolds–Orr energy equation* (Reynolds, 1895; Orr, 1907b, Art. 28). An average of the term $-\rho u_i' u_j'$ is often called the *Reynolds stress tensor* because the contraction of its product with the rate-of-strain tensor of the basic flow equals the local rate of transfer of energy from the basic flow to the perturbation.

Note that $-R^{-1} \int_{\mathcal{V}} (\partial u_i'/\partial x_j)^2 \, d\mathbf{x} < 0$ for all perturbations because viscosity always dissipates energy. Also the cubic terms have been integrated out, because the nonlinear terms in the original form of the Navier–Stokes equations are the ones which give the cubic terms here and they represent only the convection of energy, not its creation or dissipation; as a result, the energy equation is the same as it would be if the problem were linearized (because no energy enters $\mathcal{V}$ through its boundary $\partial \mathcal{V}$). It follows that $K^{-1} dK/dt$ depends on the *spatial* structure of the velocity perturbation, but not directly on its magnitude; however, the spatial structure itself does in general depend on the magnitude of the perturbation.

The *energy method* is a means of deducing rigorously conditions for stability, conditions often of some generality. Its fundamental idea is to show that if R is sufficiently small, say $R < \tilde{R}$, then the viscous dissipation is so large that surely

$$-\int_{\mathcal{V}} \left[D_{ij} u_i' u_j' + R^{-1} \left(\frac{\partial u_i'}{\partial x_j} \right)^2 \right] d\mathbf{x} < 0.$$

It follows that

$$\frac{dK}{dt} < 0.$$

Therefore $K(t) < K(0)$ for all $t > 0$, that is, the basic flow is stable in the mean. This plausibly implies that $\mathbf{u}'(\mathbf{x}, t) \to \mathbf{0}$ as $t \to \infty$ for all $\mathbf{x} \in \mathcal{V}$. However, it is conceivable that $K(t) \to 0$ as $t \to \infty$ and yet $\mathbf{u}'(\mathbf{x}_0, t)$ at *one* point $\mathbf{x}_0 \in \mathcal{V}$ does not remain small as $t \to \infty$, so the flow may be stable or unstable according to the choice of the norm in the definition of stability.

In this way it may be possible to prove a sufficient condition for stability (in the mean) to perturbations of all magnitudes, $R < \tilde{R}$, by finding $\tilde{R}$ such that

$$\frac{1}{\tilde{R}} = \sup_{\mathbf{u}'} \left[-\frac{\int_{\mathcal{V}} D_{ij} u_i' u_j' \, d\mathbf{x}}{\int_{\mathcal{V}} (\partial u_i' / \partial x_j)^2 \, d\mathbf{x}} \right] \tag{5.29}$$

for all $\mathbf{u}'$ such that $\nabla \cdot \mathbf{u}' = 0$ in $\mathcal{V}$ and $\mathbf{u}' = \mathbf{0}$ or is periodic on $\partial \mathcal{V}$ (see Exercise 5.13). Note that this argument is essentially that of the direct method of Liapounov, with K as the Liapounov functional. It follows that the basic flow is globally asymptotically stable in the mean (with norm $\|\mathbf{u}\| = (2K)^{1/2}$) if $R < \tilde{R}$, and therefore that $R_c \geq \tilde{R}$.

It is often valuable to know that a basic flow is the unique steady solution when $R < \tilde{R}$. For example, if it so happens that $\tilde{R} = R_c$, then subcritical instability cannot occur.

There has been much research on the energy method since the original work of Reynolds in 1895, with use of variational principles and so forth. Joseph (1976) applied the energy method to many problems of hydrodynamic stability. Straughan (1982) gives an integrated account of the method and its application to continuum mechanics at large as well as hydrodynamic stability. There is not space to elaborate this here, but note that if $\tilde{R}$ can be found then $\tilde{R} \leq R_c$, although it is possible that $\tilde{R}$ is so much less than R_c as to be of little practical value in indicating when the basic flow is stable and when unstable. However, *Serrin's theorem* is one such result of great theoretical value: all steady basic flows $\mathbf{U}_*(\mathbf{x})$ of an incompressible viscous fluid in a bounded domain $\mathcal{V}$ are stable if $R < 5.71$, where $R = LV/\nu$, L is the maximum diameter of $\mathcal{V}$, and $V = \sup_{\mathbf{x} \in \mathcal{V}} |\mathbf{U}_*(\mathbf{x})|$. Serrin (1959) proved this result by ingenious use of inequalities, establishing that a bounded basic flow is stable if the Reynolds number is small enough. This augments the well-known result that a steady *Stokes flow*, that is, a flow at $R = 0$, is unique and stable.

The energy method is complementary to the linear theory of stability in the sense that the linear theory may show that a given flow is unstable to some perturbations but not that the flow is stable to all, whereas the energy method may show that a flow is stable to all perturbations but not that the flow is unstable to some.

Exercises

5.1 *Dispersion relations.* Show that (i) if the dispersion function $\mathcal{D}(k, \omega) = \omega^2 - \omega_0{}^2 - v^2 k^2$ for constants $\omega_0, v > 0$, then the associated 'flow' is stable; but (ii) if $\mathcal{D}(k, \omega) = \omega^2 + \omega_0{}^2 - v^2 k^2$, then the associated 'flow' is absolutely unstable.

5.2 *Absolute and convective instability.* Consider the linear model equation

$$\frac{\partial u}{\partial t} + V \frac{\partial u}{\partial x} = \sigma u$$

for $-\infty < x < \infty$, where σ is real and $V > 0$. Note that $u = U$, where $U(x, t) = 0$ for all x, t gives a solution representing a basic state of rest. Taking normal modes with $u = \mathrm{Re}\,[\mathrm{e}^{\mathrm{i}k(x-ct)}]$, find the dispersion relation giving the complex velocity c as a function of wavenumber k, and deduce a criterion for stability of each mode.

Show that if $u(x, 0) = f(x)$, where f is differentiable everywhere and $f(x) = 0$ if $|x| > X$ for some $X > 0$, then

$$u(x, t) = \mathrm{e}^{\sigma t} f(x - Vt)$$

for $t > 0$. Deduce that the null solution $U = 0$ is stable if $\sigma \leq 0$ but convectively unstable if $\sigma > 0$.

Show, however, that if $f(x) = U_0 \operatorname{sech}^2 x$ for $-\infty < x < \infty$, then $u(x, t) \to 0$ as $t \to \infty$ for fixed x only if $\sigma < 2V$.

5.3 *Convective and absolute instability.* Given the linear model equation

$$\frac{\partial u}{\partial t} + V \frac{\partial u}{\partial x} = \sigma u + \nu \frac{\partial^2 u}{\partial x^2}$$

for $V, \nu > 0$ and real σ, show that the dispersion relation $\mathcal{D}(k, \omega) = 0$ of normal modes with $u \propto \mathrm{e}^{\mathrm{i}(kx - \omega t)}$ is specified by

$$\mathcal{D}(k, \omega) = -\mathrm{i}(\omega - kV) - \sigma + \nu k^2.$$

Deduce that the null solution is stable if $\sigma < 0$ and convectively unstable if $\sigma > 0$.

5.4 *Spatial modes for the Burgers equation.* Suppose that

$$\frac{\partial u}{\partial t} + u \frac{\partial u}{\partial x} = \nu \frac{\partial^2 u}{\partial x^2},$$

take

$$u(x, t) = U + u'(x, t)$$

for a real constant U, linearize the equation for small u', and take a spatial normal mode

$$u'(x, t) = \hat{u}e^{i(\alpha x - \omega t)}$$

for a given frequency $\omega > 0$. Hence deduce that

$$\alpha = \frac{1}{2}i(U/\nu)\left[-1 \pm \left(1 - 4i\nu\omega/U^2\right)^{1/2}\right]$$

$$= \omega/U + i\nu\omega^2/U^3 + O\left(\nu^2\omega^3/U^5\right) \quad \text{or}$$

$$-iU/\nu - \omega/U + O\left(\nu\omega^2/U^3\right) \quad \text{as } \nu\omega/U^2 \to 0.$$

Find the group velocities $c_g = \partial\omega/\partial\alpha$ of these modes for small $\nu\omega/U^2$.

5.5 *The interaction of modes due to nonlinearity.* Shows that if $u = u_1 + u_2$, where

$$u_j(x, t) = \text{Re}\,[\hat{u}_j(t)\exp(i\alpha_j x)], \quad \hat{u}_j = |\hat{u}_j|\exp(i\phi_j) \quad \text{for } j = 1, 2,$$

real constant α_j and real $\phi_j(t)$, then

$$u(x, t) = |\hat{u}_1|\cos(\alpha_1 x + \phi_1) + |\hat{u}_2|\cos(\alpha_2 x + \phi_2)$$

and

$$
\begin{aligned}
u\frac{\partial u}{\partial x} = -\frac{1}{2}\Big\{ &\alpha_1|\hat{u}_1|^2 \sin 2(\alpha_1 x + \phi_1) + \alpha_2|\hat{u}_2|^2 \sin 2(\alpha_2 x + \phi_2) \\
&+ (\alpha_1 + \alpha_2)|\hat{u}_1||\hat{u}_2| \sin[(\alpha_1 + \alpha_2)x + (\phi_1 + \phi_2)] \\
&+ (\alpha_1 - \alpha_2)|\hat{u}_1||\hat{u}_2| \sin[(\alpha_1 - \alpha_2)x + (\phi_1 - \phi_2)]\Big\}.
\end{aligned}
$$

Evaluate $u\partial u/\partial x$ similarly when

$$u_j(x, t) = \text{Re}\,[\hat{u}_j(t)\exp(i\boldsymbol{\alpha}_j \cdot \mathbf{x})], \quad \hat{u}_j = |\hat{u}_j|\exp(i\phi_j),$$

and $\boldsymbol{\alpha}_j = (\alpha_j, \beta_j, \gamma_j)$, $\mathbf{x} = (x, y, z)$.

So what modes other than the first harmonics of u_1 and u_2 would you expect nonlinear interactions to excite in fluid mechanics?

5.6 *The growth of a wave packet in space and time.* A linearized partial differential equation is given in the operational form

$$\frac{\partial u}{\partial t} + if(-i\partial/\partial x)u = 0.$$

Show that complex wave solutions $u(x, t) = e^{i(kx - \omega t)}$ have the dispersion relation $\omega = f(k)$.

Consider a wave-packet solution described by $u(x, t) = A(x, t)U(x, t)$, where the complex amplitude A is a slowly varying function of space and time, and the 'carrier wave' $U(x, t) = e^{i(Kx - \Omega t)}$ has frequency $\Omega = f(K)$ for a given wavenumber $K \neq 0$. Deduce that

$$i\left[\frac{\partial A}{\partial t} + f'(K)\frac{\partial A}{\partial x}\right] + \frac{1}{2}f''(K)\frac{\partial^2 A}{\partial x^2} = 0$$

approximately if f'' is continuous.

You are given further that the weakly nonlinear approximation for real waves leads to the equation

$$i\left[\frac{\partial A}{\partial t} + f'(K)\frac{\partial A}{\partial x}\right] + \frac{1}{2}f''(K)\frac{\partial^2 A}{\partial x^2} + l|A|^2A = 0,$$

where l is some complex Landau constant and now $u(x, t) = \mathrm{Re}[A(x, t)U(x, t)]$. Remove the term in $\partial A/\partial x$ by a Galilean transformation (noting that the group velocity $c_g = f'(K)$), and rescale x and t to deduce a *nonlinear Schrödinger equation* of the form

$$i\frac{\partial A}{\partial T} + \frac{\partial^2 A}{\partial X^2} + l|A|^2A = 0.$$

[A term $l|A|^2A$ arises in general because it is the leading approximation to the self-interaction of a weakly nonlinear wave in a homogeneous medium, homogeneity implying that wave propagation is invariant under translation; see Exercise 2.3.]

5.7 *Some solvability conditions.* (i) Define an inner product such that

$$\langle u, v \rangle = \int_0^\pi u(z)v(z)\,\mathrm{d}z$$

for all $u, v \in C^2[0, \pi]$, and a linear operator L: $C^2[0, \pi] \to C[0, \pi]$ such that

$$\mathrm{L}v = \frac{\mathrm{d}^2 v}{\mathrm{d}z^2} + p\frac{\mathrm{d}v}{\mathrm{d}z} + qv$$

for given continuously differentiable functions p, q.

Then show, by integration by parts, that

$$\langle u, \mathrm{L}v \rangle = \langle \mathrm{L}^\dagger u, v \rangle + \left[u\frac{\mathrm{d}v}{\mathrm{d}z} - v\frac{\mathrm{d}u}{\mathrm{d}z} + puv\right]_0^\pi,$$

where $\mathrm{L}^\dagger$, called the *adjoint* of L, is the linear operator defined by

$$\mathrm{L}^\dagger u = \frac{\mathrm{d}^2 u}{\mathrm{d}z^2} - \frac{\mathrm{d}(pu)}{\mathrm{d}z} + qu.$$

This is sometimes called *Lagrange's identity*. Deduce that if moreover $u, v = 0$ at $z = 0, \pi$, then

$$\langle u, \mathrm{L}v \rangle = \langle \mathrm{L}^\dagger u, v \rangle.$$

(ii) Consider the problem

$$\mathrm{L}u_n = h_n, \qquad u_n = 0 \quad \text{at } z = 0, \pi,$$

for a given continuous function h_n, where it is known that there exists a non-null function $u_1^\dagger$ such that

$$\mathrm{L}^\dagger u_1^\dagger = 0, \qquad u_1^\dagger = 0 \quad \text{at } z = 0, \pi.$$

Taking $u = u_1^\dagger$, $v = u_n$, deduce that a *necessary* condition for the existence of u_n is

$$\langle u_1^\dagger, h_n \rangle = 0,$$

and so, equivalently,

$$\int_0^\pi u_1^\dagger(z) h_n(z)\,\mathrm{d}z = 0.$$

5.8 *The Fredholm alternative.* Given a real $n \times n$ matrix $\mathbf{A}$ and column vector $\mathbf{b} \in \mathbb{R}^n$, let us seek solutions $\mathbf{x} \in \mathbb{R}^n$ of the equation

$$\mathbf{A}\mathbf{x} = \mathbf{b}. \tag{E5.1}$$

Deduce that

$$\mathbf{x} = \mathbf{A}^{-1}\mathbf{b}, \tag{E5.2}$$

if $\mathbf{A}$ is invertible.

Next consider what may happen if $\mathbf{A}$ is not invertible. Show that if $\mathbf{y} \in \mathbb{R}^n$ is an eigenvector of the transpose matrix $\mathbf{A}^\mathrm{T}$ belonging to the eigenvalue zero, that is,

$$\mathbf{A}^\mathrm{T}\mathbf{y} = \mathbf{0} \quad \text{and} \quad \mathbf{y} \neq \mathbf{0},$$

then

$$\mathbf{y}^\mathrm{T}\mathbf{b} = \mathbf{0}, \tag{E5.3}$$

that is, (E5.3) is a necessary condition for the existence of a solution of equation (E5.1).

Deduce the *Fredholm alternative*, namely, that *either* the solution **x** of equation (E5.1) is unique, *or* there exists a solution $\mathbf{z} \neq \mathbf{0}$ of $\mathbf{Az} = \mathbf{0}$; showing that in the former case **A** is invertible and zero is not an eigenvalue of $\mathbf{A}^{\mathrm{T}}$; and in the latter case **A** is not invertible, **x** exists only if **b** is orthogonal to *all* eigenvectors of $\mathbf{A}^{\mathrm{T}}$ with zero eigenvalue, and if **x** exists it is not unique. [In fact **x** exists if and only if **b** is orthogonal to all the eigenvectors with zero eigenvalue.]

Discuss the generalization of the above results for linear operator A: $H \to H$, where H is some infinite-dimensional real vector space with inner product $\langle \cdots, \cdots \rangle$, replacing the transpose $\mathbf{A}^{\mathrm{T}}$ of **A** by the *adjoint operator* $A^{\dagger}$ of A which is defined by the identity

$$\langle v, Au \rangle = \langle u, A^{\dagger}v \rangle$$

for all $u, v \in H$. [Of course, we can identify $A^{\dagger} = \mathbf{A}^{\mathrm{T}}$ for the matrix operator **A** if we choose $H = \mathbb{R}^n$ and define the inner product as $\langle u, v \rangle = \mathbf{u}^{\mathrm{T}}\mathbf{v}$. An operator A which has the Fredholm alternative property is called a *Fredholm operator*, but not all linear operators are Fredholm operators. For a counter-example, note that the creation operator C defined by $C[x_1, x_2, \ldots]^{\mathrm{T}} = [0, x_1, x_2, \ldots]^{\mathrm{T}}$ is not a Fredholm operator because $C\mathbf{x} = \mathbf{0}$ implies that $\mathbf{x} = \mathbf{0}$, although $C\mathbf{x} = [1, 0, 0, \ldots]^{\mathrm{T}}$ has no solution.]

5.9 *Adjoint operators and biorthogonality in linear algebra of finite-dimensional spaces.* Define the usual inner product by

$$\langle \mathbf{u}, \mathbf{v} \rangle = \mathbf{v}^{*\mathrm{T}}\mathbf{u}$$

for all n-vectors $\mathbf{u}, \mathbf{v} \in \mathbb{C}^n$, and deduce that

$$\langle \mathbf{u}, \mathbf{u} \rangle \geq 0, \qquad \lambda\langle \mathbf{u}, \mathbf{v} \rangle = \langle \lambda\mathbf{u}, \mathbf{v} \rangle = \langle \mathbf{u}, \lambda^*\mathbf{v} \rangle, \qquad \langle \mathbf{v}, \mathbf{u} \rangle = \langle \mathbf{u}, \mathbf{v} \rangle^*,$$

for $\lambda \in \mathbb{C}$.

Given a complex $n \times n$ matrix **A**, define the *adjoint* $\mathbf{A}^{\dagger}$ of **A** such that

$$\langle \mathbf{Au}, \mathbf{v} \rangle = \langle \mathbf{u}, \mathbf{A}^{\dagger}\mathbf{v} \rangle$$

for all $\mathbf{u}, \mathbf{v} \in \mathbb{C}^n$. Show that $\mathbf{A}^{\dagger} = \mathbf{A}^{*\mathrm{T}}$.

Show that if $\mathbf{Au} = \lambda\mathbf{u}$ for some complex eigenvector $\mathbf{u} \neq \mathbf{0}$ and eigenvalue λ, then there exists an eigenvector $\mathbf{v} \neq \mathbf{0}$ of the adjoint matrix such that $\mathbf{A}^{\dagger}\mathbf{v} = \lambda^*\mathbf{v}$.

Show further that if

$$\mathbf{A}\mathbf{u} = \lambda\mathbf{u} \quad \text{and} \quad \mathbf{A}^{\dagger}\mathbf{v} = \mu\mathbf{v}$$

for $\mathbf{u}, \mathbf{v} \neq \mathbf{0}$, then either $\mu = \lambda^*$ or

$$\langle \mathbf{u}, \mathbf{v} \rangle = 0.$$

[This property of the eigenvectors of $\mathbf{A}$ and its adjoint is called *biorthogonality*, and we often normalize so that $\langle \mathbf{u}, \mathbf{v} \rangle = 1$ if $\mu = \lambda^*$.]

5.10 *Nonlinear modes for the Burgers equation.* Given $\nu > 0$ and a continuous real function f, suppose that

$$\frac{\partial u}{\partial t} + u\frac{\partial u}{\partial x} = \nu\frac{\partial^2 u}{\partial x^2},$$

$$u(x, 0) = f(x) \quad \text{for } 0 \le x \le 2\pi,$$

$$u(2\pi, t) = u(0, t), \quad u_x(2\pi, t) = u_x(0, t) \qquad \text{for } t > 0.$$

Let $u' = u - U$ for some real constant 'basic velocity' U, and show that

$$\frac{\partial u'}{\partial t} + U\frac{\partial u'}{\partial x} + u'\frac{\partial u'}{\partial x} = \nu\frac{\partial^2 u'}{\partial x^2}. \tag{E5.4}$$

Taking the Fourier expansion

$$u'(x, t) = \sum_{n=-\infty}^{\infty} u_n(t)\mathrm{e}^{\mathrm{i}nx},$$

where $u_{-n} = u_n^*$ in order that u' is real, show that

$$u_m(0) = \frac{1}{2\pi}\int_0^{2\pi} f(x)\mathrm{e}^{-\mathrm{i}mx}\,\mathrm{d}x$$

for $m = \pm1, \pm2, \dots$. What is $u_0(0)$? By multiplying equation (E5.4) by $\mathrm{e}^{-\mathrm{i}mx}$ and integrating from $x = 0$ to 2π, show that

$$\frac{\mathrm{d}u_m}{\mathrm{d}t} = -\mathrm{i}mUu_m - \nu m^2 u_m - \frac{1}{2}\mathrm{i}m\sum_{n=-\infty}^{\infty} u_{m-n}u_n.$$

Deduce that $u_0(t) = u_0(0)$, and that if ν is large, then

$$u_1(t) \approx u_1(0)\exp[-\mathrm{i}(U + u_0)t - \nu t].$$

5.11 *Stability of a uniform flow.* Verify that a uniform basic flow of an incompressible viscous fluid with constant velocity $\mathbf{U}$ and pressure P gives an exact solution of the Navier–Stokes equations of motion and the equation of continuity. Writing $\mathbf{u}(\mathbf{x}, t) = \mathbf{U} + \mathbf{u}'(\mathbf{x}, t)$, and assuming that $\mathbf{u}'$ vanishes on the boundary $\partial \mathcal{V}$ of the domain $\mathcal{V}$ of flow (but not that $\mathbf{u}'$ is small), deduce that

$$\frac{d}{dt} \int_{\mathcal{V}} \frac{1}{2} \mathbf{u}'^2 d\mathbf{x} = -\nu \int_{\mathcal{V}} \left(\frac{\partial u_i'}{\partial x_j} \right)^2 d\mathbf{x}.$$

Deduce that the flow is stable (in the mean). Discuss the relevance of this result to Serrin's theorem on the stability of a basic flow at sufficiently small values of the Reynolds number. [See, e.g., Drazin & Reid (1981, §53.1).]

5.12 *The stability of uniformly rotating fluid.* Consider the stability of a uniform rotation of an incompressible viscous fluid within a rigid container. First show that the Navier–Stokes equations referred to a frame rotating with constant angular velocity Ω are

$$\frac{\partial \mathbf{u}}{\partial t} + \mathbf{u} \cdot \nabla \mathbf{u} + 2\Omega \times \mathbf{u} = -\nabla \left[\frac{p}{\rho} + \frac{1}{2}(\Omega \times \mathbf{x})^2 \right] + \nu \Delta \mathbf{u}, \quad \nabla \cdot \mathbf{u} = 0.$$

[Hint: see, e.g., Batchelor (1967, §3.2 and equation (7.6.1)).]

Then show that the basic flow may be taken as having zero relative velocity $\mathbf{U} = \mathbf{0}$ and pressure $P = -\frac{1}{2}\rho(\Omega \times \mathbf{x})^2$ within the domain $\mathcal{V}$, say, of the relatively stationary container $\partial \mathcal{V}$. Using Gauss's divergence theorem and the boundary condition that $\mathbf{u} = \mathbf{0}$ on $\partial \mathcal{V}$, prove that

$$\frac{d}{dt} \int_{\mathcal{V}} \frac{1}{2} u_i^2 d\mathbf{x} = -\nu \int_{\mathcal{V}} \left(\frac{\partial u_i}{\partial x_j} \right)^2 d\mathbf{x}.$$

Deduce that the flow is stable. [After Drazin & Reid (1981, Problem 1.2); Sorokin (1961, p. 372).]

5.13 *The Reynolds–Orr energy equation.* Define $\tilde{R}$ by

$$\tilde{R}^{-1} = \sup_{u_i' \in S} \left[-\int_{\mathcal{V}} D_{ij} u_i' u_j' \, d\mathbf{x} \Big/ \int_{\mathcal{V}} \left(\frac{\partial u_i'}{\partial x_j} \right)^2 d\mathbf{x} \right], \qquad (E5.5)$$

where S is the set of continuously differentiable vector fields u_i' over $\mathcal{V}$ such that

$$\nabla \cdot \mathbf{u}' = 0 \quad \text{in } \mathcal{V} \quad \text{and} \quad \mathbf{u}' = \mathbf{0} \quad \text{on } \partial \mathcal{V}. \qquad (E5.6)$$

Show that

$$\frac{\mathrm{d}K}{\mathrm{d}t} = \left[-\int_{\mathcal{V}} D_{ij}u_i'u_j'\, \mathrm{d}\mathbf{x} \Big/ \int_{\mathcal{V}} \left(\frac{\partial u_i'}{\partial x_j}\right)^2 \mathrm{d}\mathbf{x} - R^{-1} \right] \int_{\mathcal{V}} \left(\frac{\partial u_i'}{\partial x_j}\right)^2 \mathrm{d}\mathbf{x},$$

where $K = \int_{\mathcal{V}} \frac{1}{2}u_i'^2\, \mathrm{d}\mathbf{x}$, and deduce that

$$\frac{\mathrm{d}K}{\mathrm{d}t} \le \left(\tilde{R}^{-1} - R^{-1}\right) \int_{\mathcal{V}} \left(\frac{\partial u_i'}{\partial x_j}\right)^2 \mathrm{d}\mathbf{x},$$

$$\le 0 \quad \text{if } R < \tilde{R}.$$

Show that the variational principle derived from equation (5.29) gives Euler–Lagrange equations

$$D_{ij}u_j' = -\frac{\partial \lambda'}{\partial x_i} + \rho^{-1}\Delta u_i', \tag{E5.7}$$

where $-2\lambda'$, ρ^{-1} are Lagrange multipliers associated with the respective constraints that the divergence of the velocity is zero and that the dissipation integral is normalized. Show that if R is less than the least eigenvalue ρ of the *linear* problem (E5.7), (E5.6), then the basic flow is globally asymptotically stable, that is, asymptotically stable to all perturbations, whatever their magnitude.

6

Rayleigh–Bénard Convection

When you see a cloud rise out of the west, straightway ye say, There cometh a shower; and so it is.

Luke xii 54

The treatment of particular problems of instability of flows is resumed in this chapter. The instability when a fluid is heated from below is manifest as thermal convection. This is modelled by a classic problem which is solved mathematically and related to observations. The problem is important as a prototype of thermal convection and of transition to turbulence.

6.1 Thermal Convection

In 1900 Bénard made some quantitative experiments on thermal convection. He melted some wax in a metal dish by heating the base, there being a layer of wax about 1 mm deep. When the base was hot enough to melt all the wax, there was at first no motion of the liquid wax. But as the base was heated above some critical temperature, Bénard saw a hexagonal pattern develop on the surface of the wax, and deduced the presence of convection cells below. Look at Figure 6.1 to see what he saw.

Rayleigh modelled this problem in 1916, and treated it by use of the theory of hydrodynamic stability. He assumed that there was an infinite layer of a fluid bounded by stationary horizontal planes, $z_* = 0$ and $z_* = d$, say, which are maintained at constant uniform temperatures, θ_0, θ_1 respectively. The configuration and notation are shown in Figure 6.2.

There is a basic state of rest, with temperature governed by conduction and pressure in hydrostatic balance. This gives basic velocity, temperature and pressure as

$$\mathbf{U}_* = \mathbf{0}, \qquad \Theta_* = \theta_0 - \beta z_*, \qquad P_* = p_0 - g\rho_0\left(z_* + \tfrac{1}{2}\alpha\beta z_*^2\right), \quad (6.1)$$

respectively, for $0 \leq z_* \leq d$, where $\beta = (\theta_0 - \theta_1)/d$ is the basic adverse temperature gradient, ρ_0 is the mean density, and α is the coefficient of cubical expansion of the fluid. We anticipate that if β is small enough, then viscosity and thermal diffusion will stabilize the flow, even when $\beta > 0$ (i.e. hot fluid

93

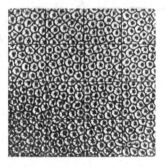

Figure 6.1   Plan view of the surface of a layer of spermaceti wax heated from below. After Bénard (1900).

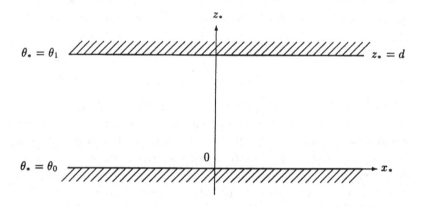

Figure 6.2   The configuration of Rayleigh–Bénard convection.

is below cooler and therefore denser fluid), but that if β exceeds some critical value, then an 'overturning' instability will ensue.

Boussinesq (1903, vol. II, p. 172), and, in fact, Oberbeck (1879) before him, had recognized that when temperature variations are small, the variations of the thermodynamic properties, such as viscosity, thermal diffusivity, density and specific heat, of a fluid are small and the fluid is approximately incompressible, although the buoyancy of the fluid is significant. This is because the acceleration of the fluid is much less than the acceleration of gravity, yet the product of g and a small density difference may not be negligible relative to some other terms in the vertical equation of motion. Also they assumed an equation of state in which the density is a linear function of temperature and independent of pressure. In short, Rayleigh chose to model the convection in a thin layer by using their equations of motion, energy and state for a *Boussinesq fluid*, which are respectively

$$\frac{\partial \mathbf{u}_*}{\partial t_*} + \mathbf{u}_* \cdot \nabla_* \mathbf{u}_* = -\nabla_* (p_*/\rho_0 + g z_*) + \alpha g(\theta_* - \theta_0)\mathbf{k} + \nu \Delta_* \mathbf{u}_*, \quad (6.2)$$

$$\nabla_* \cdot \mathbf{u}_* = 0, \quad (6.3)$$

$$\frac{\partial \theta_*}{\partial t_*} + \mathbf{u}_* \cdot \nabla_* \theta_* = \kappa \Delta_* \theta_*, \quad (6.4)$$

$$\rho_* = \rho_0 \left[1 - \alpha(\theta_* - \theta_0) \right]. \quad (6.5)$$

We use asterisks as subscripts to denote dimensional forms of the dependent and independent variables. These are called the *Boussinesq equations*.

At a stationary rigid surface, say $z_* = $ constant, there is no slip and no penetration of the fluid so that

$$\mathbf{u}_* = \mathbf{0}.$$

This models a flat horizontal plate. Rayleigh modelled a free surface as a stationary horizontal one with zero *lateral* stress, so that

$$w_* = \frac{\partial u_*}{\partial z_*} + \frac{\partial w_*}{\partial x_*} = \frac{\partial w_*}{\partial y_*} + \frac{\partial v_*}{\partial z_*} = 0$$

at a free surface with equation $z_* = $ constant. If the surface is a perfect conductor, then, moreover,

$$\theta_* = \text{constant}.$$

Other boundary conditions, for example those for perfect insulators when the basic temperature gradient is due to an internal heat source, also have been considered.

You can verify that the basic state of rest is a solution of the above equations.

6.2 The Linearized Problem

We write

$$\mathbf{u}_* = \mathbf{u}'_*(\mathbf{x}_*, t_*), \quad \theta_* = \Theta_*(z_*) + \theta'_*(\mathbf{x}_*, t_*), \quad p_* = P_*(z_*) + p'_*(\mathbf{x}_*, t_*);$$

linearize the Boussinesq equations (6.2)–(6.5) for small perturbations; define dimensionless variables $\mathbf{x} = \mathbf{x}_*/d$, $t = \kappa t_*/d^2$, $\mathbf{u} = d\mathbf{u}'_*/\kappa$, $\theta = \theta'_*/\beta d$,

$p = d^2 p'/\rho_0 \kappa^2$; and deduce that

$$\frac{\partial \mathbf{u}}{\partial t} = -\nabla p + RPr\theta \mathbf{k} + Pr\Delta \mathbf{u}, \qquad (6.6)$$

$$\nabla \cdot \mathbf{u} = 0, \qquad (6.7)$$

$$\frac{\partial \theta}{\partial t} - w = \Delta \theta, \qquad (6.8)$$

where $R = \alpha \beta g d^4/\kappa \nu$ is called the *Rayleigh number*, and $Pr = \nu/\kappa$ is the *Prandtl number*. The Rayleigh number is a dimensionless measure of the ratio of the destabilizing effect of the buoyancy (for $\beta > 0$) to the stabilizing effect of molecular diffusion of momentum and buoyancy, and Pr is a property of the fluid (not the configuration of the layer at all).

The curl of the equation (6.6) gives

$$\frac{\partial \boldsymbol{\omega}}{\partial t} = RPr\nabla\theta \times \mathbf{k} + Pr\Delta \boldsymbol{\omega}, \qquad (6.9)$$

where $\boldsymbol{\omega} = \nabla \times \mathbf{u}$ is the vorticity. The curl of equation (6.9) gives

$$\frac{\partial(\Delta \mathbf{u})}{\partial t} = RPr\left(\Delta\theta \mathbf{k} - \nabla\frac{\partial\theta}{\partial z}\right) + Pr\Delta^2\mathbf{u},$$

on use of the continuity equation (6.7). The z-component of this equation is

$$\frac{\partial(\Delta w)}{\partial t} = RPr\Delta_1\theta + Pr\Delta^2 w, \qquad (6.10)$$

where $\Delta_1 = \partial^2/\partial x^2 + \partial^2/\partial y^2$ is the horizontal Laplacian. Elimination of θ from (6.8) and (6.10) finally gives

$$\left(\frac{\partial}{\partial t} - \Delta\right)\left(\frac{1}{Pr}\frac{\partial}{\partial t} - \Delta\right)\Delta w = R\Delta_1 w. \qquad (6.11)$$

The boundary conditions are already linear. They, *and* the continuity equation (6.7), give

$$w = \frac{\partial w}{\partial z} = \theta = 0$$

at a rigid plate $z = $ constant, and

$$w = \frac{\partial^2 w}{\partial z^2} = \theta = 0$$

at a free surface $z = $ constant.

Rayleigh took normal modes of the form

$$\theta = T(z)f(x, y)e^{st}, \qquad w = W(z)f(x, y)e^{st}. \tag{6.12}$$

In order that the variables are separable in this way, equation (6.8) gives Helmholtz's equation for f,

$$\Delta_1 f + a^2 f = 0, \tag{6.13}$$

for some constant a^2 of separation; we identify a as the horizontal wavenumber of the mode. Then equations (6.8), (6.10) become

$$\left(D^2 - a^2 - s\right)T = -W, \tag{6.14}$$

$$\left(D^2 - a^2\right)\left(D^2 - a^2 - s/Pr\right)W = a^2 RT, \tag{6.15}$$

respectively; equation (6.11) becomes

$$\left(D^2 - a^2\right)\left(D^2 - a^2 - s\right)\left(D^2 - a^2 - s/Pr\right)W = -a^2 RW; \tag{6.16}$$

and the boundary conditions at a perfect conductor become

$$\left.\begin{array}{l} W = DW = T = 0 \text{ (rigid)} \\ W = D^2 W = T = 0 \text{ (free)} \end{array}\right\} \quad \text{at } z = 0, 1; \tag{6.17}$$

where $D = d/dz$ and $T = (a^2 R)^{-1}(D^2 - a^2)(D^2 - a^2 - s/Pr)W$. Equations (6.16), (6.17) are the eigenvalue problem to determine the linear stability characteristics.

6.3 The Stability Characteristics

It can be proved quite generally (see Exercise 6.2) that the principle of the exchange of stabilities is valid for these problems, that is, if $R < 0$, then $\text{Re}(s) < 0$ and so the flow is stable, and if $R > 0$, then $\text{Im}(s) = 0$. The first result is physically plausible – if the lower plate is cooled (and so $\theta_0 < \theta_1$), then there is stability. The second result implies that $s = 0$ at the margin of stability (wherever that is).

However, Rayleigh (1916a) was able to solve the problem explicitly in the special case of *two* free boundaries at $z = 0, 1$, as in Example 6.1 below. These are sometimes called *free–free boundaries*. He chose both boundaries to be free for mathematical convenience; although a free boundary at the bottom may seem artificial, it can be simulated by replacing the bottom plate by a layer of a much less viscous liquid (Goldstein & Graham, 1969). The explicit solution is also useful in explaining the structure of the solutions to the problem with other boundary conditions.

Example 6.1: The stability characteristics for free–free perfectly conducting boundaries. Here we need to solve equation (6.16) with free–free boundary conditions (6.17). It can be seen by inspection that

$$W = W_n, \quad s = s_n \qquad \text{for } n = 1, 2, \ldots, \tag{6.18}$$

where

$$W_n(z) = \sin n\pi z \tag{6.19}$$

and

$$(n^2\pi^2 + a^2)(n^2\pi^2 + a^2 + s_n)(n^2\pi^2 + a^2 + s_n/Pr) = a^2 R. \tag{6.20}$$

Therefore

$$s_n = -\tfrac{1}{2}(1 + Pr)(n^2\pi^2 + a^2)$$
$$\pm \left[\tfrac{1}{4}(Pr - 1)^2(n^2\pi^2 + a^2)^2 + a^2 R Pr/(n^2\pi^2 + a^2)\right]^{1/2}.$$

We can now verify in this case that $R < 0$ implies that $\operatorname{Re}(s_n) < 0$ and $R > 0$ implies that $s_n = 0$ at the margin of stability. So let us simply put $s_n = 0$ in the quadratic for s_n and find the value of R for marginal stability of the nth mode of wavenumber a,

$$R_n(a) = (n^2\pi^2 + a^2)^3/a^2. \tag{6.21}$$

The graph of $R_1(a)$ is sketched in Figure 6.3. It is called a *marginal curve*, giving the boundary between growing and decaying modes in the plane of the Rayleigh number and the wavenumber, such that the minimum value of the Rayleigh number on the curve is its critical value.

To find the margin of the stability of the flow, we find the minimum of R_n for all n by putting $n = 1$ (for the least stable vertical mode), and then the minimum for all real horizontal wavenumbers a. This gives

$$R_c = \min_{-\infty < a < \infty} R_1(a^2) = 27\pi^4/4$$

$$= 657.5,$$

the minimum occurring when $a = a_c = \pi/2^{1/2} = 2.221$. It follows that if $R < R_c$ there is stability and if $R > R_c$ there is instability. If R were slightly above R_c, then we would expect wave modes with $n = 1$ and $a \approx a_c$, but no others, to grow exponentially. □

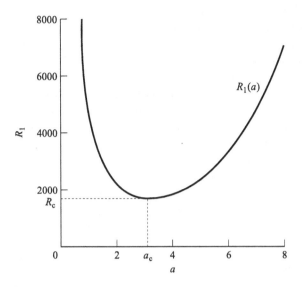

Figure 6.3 The marginal curve: graph of the critical Rayleigh number R_1 against the dimensionless wavenumber a. Note the minimum of R_1 at $a = a_c$. (Adapted from Drazin & Reid, 1981, Fig. 2.2(a).)

Table 6.1. *The critical values of the Rayleigh number and wavenumber for Rayleigh–Bénard convection between perfectly conducting horizontal planes, with various boundary conditions*

	Free–free	Free–rigid	Rigid–rigid
R_c	657.5	1101	1708
a_c	2.221	2.682	3.117

This type of instability is sometimes called *convective instability*, sometimes Rayleigh–Bénard instability, sometimes *Rayleigh–Bénard convection*. Computation is needed to find R_c and a_c for convection in the cases of two rigid plates, sometimes called rigid–rigid, and of a rigid bottom and free top surface, sometimes called free–rigid (see Table 6.1). The shapes of the curves $R = R_n(a)$ are similar qualitatively to the shape of the curve $R = R_1(a)$ in Figure 6.3 in each case.

We have determined the size of the cells via the horizontal wavenumber a_c, but the shape of the cells via the solution f of Helmholtz's equation (6.13) is not determinate in *this* model. Modelling a thin layer of fluid by an unbounded layer, we seek solutions of Helmholtz's equation which tessellate the plane with periodic cells, for example, triangular, rectangular, parallelogram or hexagonal cells. For more realism, we need side walls (which render the wavenumber a discrete) or nonlinearity (which can render some cell shapes unstable) to find the cell shapes via the function f. Then, for example, if the walls have a circular section, we may expect annular cells (see Van Dyke, 1982, Fig. 140).

The simplest case is of long rolls. It can be seen by inspection that $f(x, y) = \cos ax$ is a solution of equation (6.13), and has period $2\pi/a$ in x. Also it can be shown from equation (6.7) of continuity that in this case

$$u = -a^{-1} \sin ax\, \mathrm{D}W \mathrm{e}^{st}, \qquad v = 0, \qquad w = \cos ax\, W \mathrm{e}^{st},$$

and therefore $u = 0$ at $x = k\pi/a$ for $k = 0, \pm 1, \pm 2, \ldots$. It follows that this solution does indeed represent long roll cells parallel to the y-axis.

In the theory of Rayleigh which we have described, there is an infinite layer of fluid, so that the wavenumber a is taken as a continuous parameter. Any real layer of fluid is, however, bounded – there must be side walls to confine the fluid – and an infinite layer is regarded only as an approximation to a layer of large horizontal dimensions. So it is better to take side walls (see Exercise 6.7); this renders the wavenumber an eigenvalue, and hence a *discrete* variable, for given Rayleigh number and domain of flow. For a layer whose depth is comparable to its width, the discreteness of the eigenvalues and their eigenfunctions is physically important.

It is time to look at convection again. Examine Van Dyke's (1982, Figs. 139–142) book for a start. Look at Figure 6.4 to see the measured values of the heat transfer (see Exercise 6.8) as a function of the Rayleigh number for fluids of various values of the Prandtl number. The *Nusselt number*, that is, the ratio of the actual heat transfer to what the heat transfer would be if there were pure conduction of heat and no convection, is plotted against the Rayleigh number. Thus the Nusselt number is 1 in the absence of convection. Note how the value of the Rayleigh number at the onset of instability is found to be almost independent of the fluid and hence of the Prandtl number. You may with advantage make your own experiment (see Exercise 6.18).

6.4 Nonlinear Convection

In summary, the linear theory gives the most unstable mode with

$$\theta(\mathbf{x}, t) = A(t) f(x, y, a) T_1(z), \tag{6.22}$$

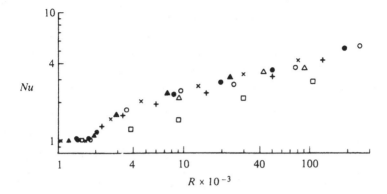

Figure 6.4   Some experimental results on the heat transfer in various fluids in various containers. The Nusselt number is plotted against the Rayleigh number: ○, water ($Pr = 7.02$); +, heptane ($Pr = 7.05$); ×, ethylene glycol; ●, silicone oil AK 3 ($Pr = 4.20$); ▲, silicone oil AK 350 ($Pr = 44.0$); △, air ($Pr = 0.62$); □, mercury ($Pr = 0.025$). After Drazin & Reid (1981, Fig. 2.6), Silveston (1958) and Rossby (1969).

where

$$\Delta_1 f + a^2 f = 0,$$
$$\frac{\mathrm{d}A}{\mathrm{d}t} = s_1 A,$$

and

$$s_1 \sim k(R - R_c) \quad \text{as } R \to R_c$$

for some $k > 0$. For free–free convection, $T_1(z) = \sin \pi z$. It is plausible that if $R > R_c$ but $R - R_c$ is small enough, then there is only one unstable mode (strictly, we need to model the side walls to render the horizontal wavenumber a discrete to justify this). Then all other modes will decay exponentially, and so may be neglected. However, the unstable mode, starting with small amplitude $A(0)$, grows slowly with a small exponential rate $s \sim k(R - R_c)$. Eventually the nonlinearity will become significant and moderate the exponential growth (the smaller $R - R_c$ is, the longer this will take).

The weakly nonlinear theory shows essentially that, for only one unstable mode,

$$\frac{\mathrm{d}A}{\mathrm{d}t} = sA - l_1|A|^2 A - l_2|A|^4 A - \cdots$$

for a general wave mode of a problem invariant under translation. (This special form of amplitude equation arises for modes which are independent of the *phase* of the complex amplitude; if a wave with $Af = |A|e^{i(\eta+ax)}$ is invariant under translations in the x-direction, the amplitude evolution equation must be independent of changes of η, which correspond to changes of kx.) In Rayleigh–Bénard convection we have found real s and may deduce that there is no need for complex numbers, at least for roll cells, so we may take $A, f, l_1, l_2, \ldots$ as real. So

$$\frac{dA}{dt} = sA - l_1 A^3 - l_2 A^5 - \cdots .$$

In general, l_1 may be positive or negative. But calculations for Rayleigh–Bénard convection in fact give, by application of perturbation theory as in Example 5.3, $l_1 > 0$. This means that there is equilibration, and we may rationally and consistently neglect $l_2 A^5, \ldots$ if A_0 is small and $0 < R - R_c \ll 1$. We deduce the canonical equation for a supercritical pitchfork bifurcation, much as in §5.2,

$$\frac{dA}{dt} = k(R - R_c)A - l_1 A^3. \tag{6.23}$$

The 'handle' and the 'middle prong' of the pitchfork depict the solution $A = 0$, which represents the basic state of rest. The other two 'prongs' depict the solutions $A = \pm[k(R - R_c)/l_1]^{1/2}$, which represents cellular motions, $+$ for flow in one direction and $-$ for flow in the opposite direction. We can calculate the magnitude of the velocity and temperature fields from this, whereas the linear theory gives only the spatial structure of the cellular motion and its (transient) rate of growth. Note that the nonlinear theory gives the cellular motion irrespective of the initial perturbation, which determines only the direction of the flow eventually.

The nature of the primary bifurcation, from the state of rest of a thin layer of fluid heated from below as the Rayleigh number increases, depends little on the Prandtl number. After the primary bifurcation there are many further bifurcations as the Rayleigh number increases, according to the value of the Prandtl number (and other details). The pattern of bifurcations is complicated, but, at the risk of over-simplification, a useful impression may be gained from Figure 6.5, which describes compactly the results of many laboratory experiments. However, there is often hysteresis when the Rayleigh number is slowly increased and then decreased. At high values of the Rayleigh number a great variety of cell patterns have been observed (see Figure 6.6).

Most of the early experiments were on shallow layers of fluid in order to conform to Rayleigh's model. But Threlfall (1975) initiated experiments with liquid helium in containers which are not shallow. The liquid helium makes it less difficult to make accurate measurements. The cell's not being shallow

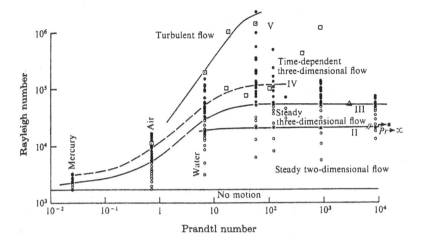

Figure 6.5 The regime diagram of the observed patterns of Rayleigh–Bénard convection in a thin layer of fluid between rigid horizontal plates for given values of the Rayleigh and Prandtl numbers. The horizontal line $Nu = 1$ marks the onset of the steady primary instability as the Rayleigh number increases. Curve II marks the secondary instability, such that the steady long rolls (with axes parallel to the shorter sides of a rectangular container) are usually preferred below the curve and steady three-dimensional cells above it. Above curve III the flow becomes unsteady. At curve IV the slope of the Nusselt number versus the Rayleigh number increases (fairly abruptly) again. In region V above curve IV, the frequencies of the unsteady flows increase as the flows becomes chaotic. ○, steady flows; •, unsteady flows; ⋆, transition points with observed change in slope; □, Rossby's (1969) observations of unsteady flow; ▣, Willis & Deardorff's (1967) observations of turbulent flow; △, Silveston's (1958) point of transition for unsteady flow. (After Krishnamurti, 1973, Fig. 4.)

reduces the rate of onset of successive bifurcations, so that they may be studied more easily, and also changes the nature of the bifurcations. Exploiting this development, Ahlers and Libchaber and their colleagues made many interesting experiments in the late 1970s and the 1980s on the transition of convection to turbulence. Of special importance is the observation of period-doubling bifurcations and onset of chaos for convection in a deep layer of fluid in a cell by Libchaber & Maurer (1978) and the theoretical interpretation of their results by Feigenbaum (1980).

There is a lot more which has been written about cell shapes, side walls, non-Boussinesq fluids, transition to turbulence as R increases far above R_c, and experimental results. Rayleigh–Bénard convection may be generalized to take into account the Marangoni effect where the surface tension of the free surface varies with temperature (see Exercise 6.17), flow in a porous medium such as sand or rock (see Exercise 6.13), double-diffusive convection when density differences are due to a solute as well as temperature (see Exercise 6.16), rotation of the basic frame of reference, radiation of heat, the Soret and Gibbs–Thomson effects in crystal growth and other applications. In fact, the Marangoni effect was important in Bénard's original experiments. Also different geometric configurations have been studied, for example a layer of fluid on a heated sphere, rotating and not rotating, with and without self-gravity. Koschmieder's (1993) book gives a lot of the details of some of these, and other, points, relating the linear and nonlinear theories to the laboratory observations. Again, Rayleigh–Bénard instability is related to many varieties of natural convection, even though the model may not represent the details of the convection very faithfully; for example, the granulation of the sun and clouds is shown in Figure 6.7. Further, much of the theory of Rayleigh–Bénard instability is of interest because of its wide applications in nonlinear physics; many other problems of evolution in

(a)

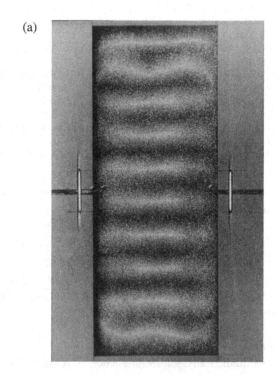

Figure 6.6   Photographs of various types of cells, steady and unsteady, which have been observed for various values of the Rayleigh and Prandtl numbers. (a) A plan view of roll cells (steady) (after Srulijes, 1979; see Koschmieder, 1993, Fig. 5.3).

(b)

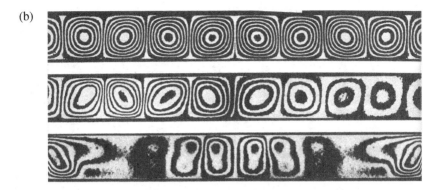

Figure 6.6 (*continued*)   (b) A side view of roll cells (steady) in a silicone oil (after Oertel & Kirchartz, 1979; see Van Dyke, 1982, Fig. 139; reproduced by permission of Springer-Verlag GmbH & Co. KG).

(c)

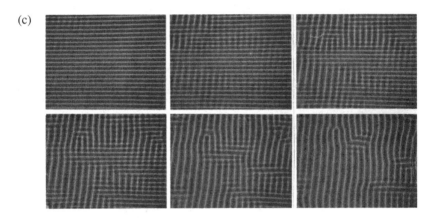

Figure 6.6 (*continued*)   (c) Cross-roll instability (steady, interaction of two rolls with the same wavelength) for $R = 3000$, $Pr = 100$, Dow Corning 200 silicone oil (after Busse & Whitehead, 1971, Fig. 10).

layers, for example, electrically driven motion in a liquid crystal, have the same symmetries and so exhibit many of the same types of instabilities and pattern formation. Perhaps Rayleigh–Bénard convection has been the subject of more theoretical research than its direct physical importance justifies, but it may be regarded as a prototype of many sorts of pattern formation and instabilities as well as convection.

(d)

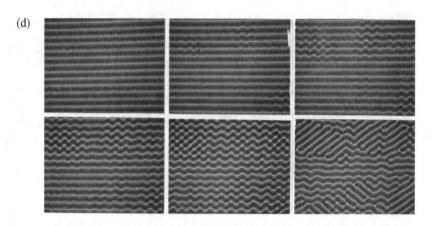

Figure 6.6 (*continued*)   (d) Zig-zag instability (steady) for $R = 3600$, $Pr = 100$ (after Busse & Whitehead, 1971, Fig. 11).

(e)

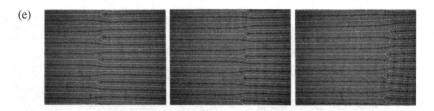

Figure 6.6 (*continued*)   (e) Pinching instability, $R = 18 \times 10^3$, $Pr = 100$ (after Busse & Whitehead, 1971, Fig. 15).

Exercises

6.1 *A linear ordinary-differential system which crudely models Rayleigh–Bénard instability.* Given, as a toy problem, that

$$\frac{dw}{dt} = -\nu d^{-2}w + \alpha g\theta, \qquad \frac{d\theta}{dt} = -\kappa d^{-2}\theta + \beta w,$$

show that there are normal modes with $w, \theta \propto e^{st}$, where

$$s = -\tfrac{1}{2}(\nu + \kappa)d^{-2} \pm \tfrac{1}{2}[(\nu - \kappa)^2 d^{-4} + 4\alpha\beta g]^{1/2}.$$

Deduce that the null solution $w = 0, \theta = 0$ is stable if $R > 1$, where $R = \alpha\beta g d^4/\kappa\nu$.

(f)

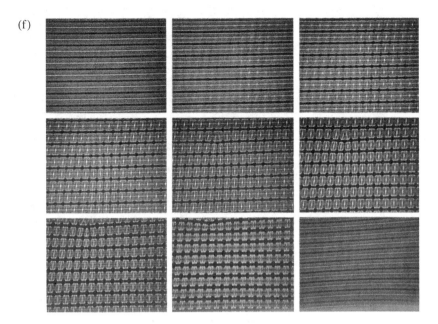

Figure 6.6 (*continued*)   (f) Bimodal instability (steady, interaction of two rolls with different wavelengths) for $R \approx 20 \times 10^3 - 65 \times 10^3$, $Pr = 100$ (after Busse & Whitehead, 1971, Fig. 14).

Show that if there is no dissipation, so that $\nu = \kappa = 0$, then

$$\frac{\mathrm{d}^2 w}{\mathrm{d}t^2} + N^2 w = 0,$$

where $N^2 = -\alpha \beta g$ defines a 'buoyancy' frequency N.

6.2   *The principle of the exchange of stabilities for Rayleigh–Bénard convection.* Multiplying equation (6.14) by the complex conjugate T^* of T, integrating from $z = 0$ to 1, and using either of the conditions (6.17) at $z = 0, 1$ for perfectly conducting or perfectly insulating stationary boundaries, show that

$$s I_0 + I_1 = \int_0^1 W T^* \, \mathrm{d}z,$$

where $I_0 = \int_0^1 |T|^2 \, \mathrm{d}z$, $I_1 = \int_0^1 (|\mathrm{D}T|^2 + a^2 |T|^2) \, \mathrm{d}z$. Multiplying equation (6.15) by W^*, show similarly that

$$J_2 + s J_1 / Pr = a^2 R \int_0^1 W^* T \, \mathrm{d}z,$$

(g)

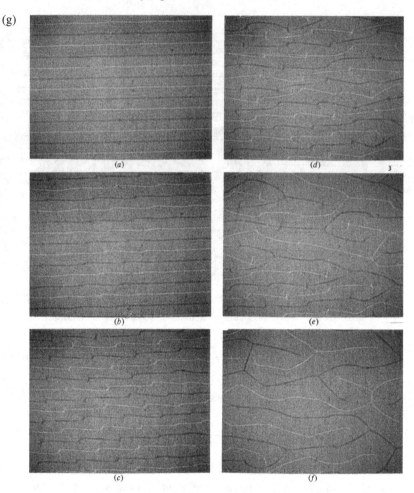

Figure 6.6 (*continued*) (g) Skewed varicose instability (unsteady) for $R \approx 10^4$, $Pr = 3.7$, water (after Busse & Clever, 1979, Fig. 7).

where $J_1 = \int_0^1 (|DW|^2 + a^2|W|^2)\, dz$ and $J_2 = \int_0^1 (|D^2W|^2 + 2a^2|DW|^2 + a^4|W|^2)\, dz$.

Hence show that if $\sigma = \mathrm{Re}(s)$, $\omega = \mathrm{Im}(s)$, then

$$\sigma\left(a^2 R I_0 - J_1/Pr\right) + a^2 R I_1 - J_2 = 0, \qquad \omega\left(a^2 R I_0 + J_1/Pr\right) = 0.$$

Deduce that if $R < 0$, then $\sigma < 0$, so there is stability, and that if $R > 0$, then $\omega = 0$, and therefore the principle of the exchange of stabilities is valid. [Pellew & Southwell (1940).]

(h)

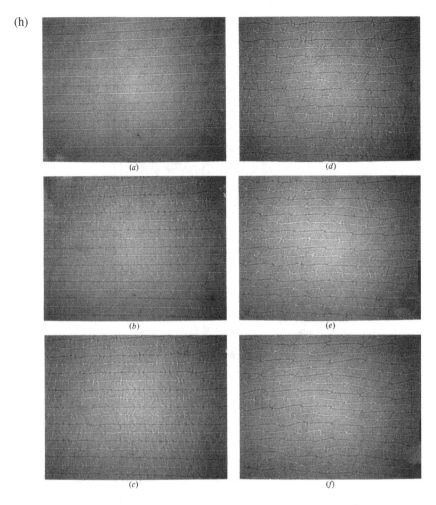

Figure 6.6 (*continued*)   (h) Knot instability (unsteady) for $R \approx 5 \times 10^4$, $Pr = 7.1$, methyl alcohol (after Busse & Clever, 1979, Fig. 8).

6.3   *Rayleigh–Bénard convection in an unusual fluid.* (i) The fluid 'Hupnol', which has density ρ_0 at a temperature θ_0, has the unusual property that perturbations of its density about ρ_0 vary as the *cube* of perturbations of its temperature from θ_0, with smaller density corresponding to larger temperature, so that $\rho = \rho_0[1 - \alpha(\theta - \theta_0)^3]$, where α is a positive constant.

   In a laboratory experiment some Hupnol is confined between two horizontal planes, which apply, to a good degree of approximation, stress-free boundary conditions to the Hupnol. The temperature is held at

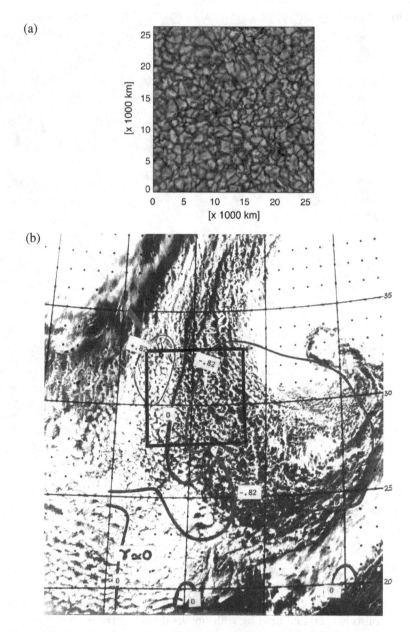

Figure 6.7   (a) Granules on the surface of the sun. The width of a convection cell is of the order of 100 km. (Reproduced by permission of the National Optical Astronomy Observatory, T. Rimmele/NOAO/AURA/NSF.) (b) Satellite photograph showing open convection cells in a region of downward motion and closed cells in a region of upward motion. (After Krishnamurti, 1975, Fig. 3(e).)

θ_0 on the lower plane $z = 0$ and θ_1 on the upper plane $z = d$. The distance d between the planes is much smaller than their horizontal extent.

Neglecting the effects of compressibility, the variation of the density on the inertia, and assuming that the coefficients of viscosity, thermal conductivity, etc., are constant, show that the stability of the state of rest is governed by the linearized problem

$$\left(\frac{\partial}{\partial t} - \kappa \Delta\right)\left(\frac{\partial}{\partial t} - \nu \Delta\right)\Delta\theta' = 3g\alpha\left(\frac{\theta_0 - \theta_1}{d}\right)^3 z^2 \Delta_1 \theta'$$

$$\theta' = \Delta\theta' = \Delta^2\theta' = 0 \quad \text{at } z = 0, d,$$

where θ' is the temperature perturbation, Δ_1 is the horizontal Laplacian operator, g is the acceleration due to gravity, κ is the thermal diffusivity, and ν is the kinematic viscosity of Hupnol.

(ii) Derive an ordinary-differential eigenvalue problem for the normal modes.

(iii) Prove that the principle of exchange of stabilities is valid for the present eigenvalue problem. [Hint: Express $\int_0^1 z^2 |T|^2 \, dz$ as a quadratic polynomial of s, the coefficients involving positive definite integrals.]

(iv) Devise a method to find an *approximate* value of the critical 'Rayleigh' number $A = g\alpha(\theta_0 - \theta_1)^3 d^3/\kappa/\nu$ below which no perturbation will grow. Thereby approximate the critical value. Describe briefly, without performing any explicit calculation, how a more accurate estimate of the critical value could be obtained. Would you expect the critical value to be greater than, equal to or less than the critical value of the Rayleigh number if water were to replace the Hupnol in the experiment? Why? [After H. E. Huppert (private communication).]

6.4 *Horizontal motion of a normal mode in Rayleigh–Bénard convection.* Use equation (6.9) to deduce the diffusion equation

$$\frac{\partial \zeta}{\partial t} = Pr \Delta \zeta, \tag{E6.1}$$

where $\zeta = \partial v/\partial x - \partial u/\partial y$ is the vertical component of the vorticity of the perturbation. Hence show (plausibly, at least) that $\zeta \to 0$ as $t \to \infty$ for fixed $\mathbf{x}$.

Now, *assuming* that $\partial v/\partial x = \partial u/\partial y$, and using the equation of continuity (6.7), show that

$$\Delta_1 u = -\frac{\partial^2 w}{\partial x \partial z}, \qquad \Delta_1 v = -\frac{\partial^2 w}{\partial y \partial z}. \tag{E6.2}$$

Deduce that, for a normal mode,

$$u = \frac{1}{a^2}\frac{\partial f}{\partial x}(DW)e^{st}, \qquad v = \frac{1}{a^2}\frac{\partial f}{\partial y}(DW)e^{st}. \qquad (E6.3)$$

6.5 *Rectangular cells.* Verify that

$$f(x, y) = \cos a_1 x \cos a_2 y$$

is a solution of Helmholtz's equation (6.13) if $a_1^2 + a_2^2 = a^2$. Show that f has period $2\pi/a_1$ in x, and period $2\pi/a_2$ in y. Show that $u = 0$ on $x = 0, \pi/a_1$ and $v = 0$ on $y = 0, \pi/a_2$.

Deduce that this function f describes motion confined in periodic cells with vertical boundaries and rectangular cross-section, each rectangle having sides of lengths $2\pi/a_1, 2\pi/a_2$.

6.6 *Hexagonal cells.* Verify that

$$f(x, y) = \cos\left[\tfrac{1}{2}a\left(3^{1/2}x + y\right)\right] + \cos\left[\tfrac{1}{2}a\left(3^{1/2}x - y\right)\right] + \cos ay$$

is a solution of Helmholtz's equation (6.13). Show that f is invariant under rotation of angle $60°$ about the z-axis, and that it gives $u = 0$ on $x = \tfrac{1}{3}3^{1/2}L, -\tfrac{1}{2}L < y < \tfrac{1}{2}L$, where $L = 4\pi/3a$.

Deduce that this function f describes motion in regular hexagonal cells, each side of each hexagon having length L. [Christopherson (1940).]

6.7 *Roll cells with side walls.* Example 6.1 gives that, for marginal stability of the first mode of free–free convection,

$$\left(a^2 + \pi^2\right)^3 = a^2 R;$$

show that this equation has three roots $a_1^2(R), a_2^2(R), a_3^2(R)$, say, where $a_1^2 > a_2^2 > 0 > a_3^2$ if $R > R_c = 27\pi^4/4$.

Hence show that if there is *steady* two-dimensional linear convection with perfectly conducting walls at $x = \pm L$, then some eigenfunctions are of the form

$$w(x, z) = \left(\frac{A_1 \cos a_1 x}{\cos a_1 L} + \frac{A_2 \cos a_2 x}{\cos a_2 L} + \frac{A_3 \cosh |a_3| x}{\cosh |a_3| L}\right)\sin \pi z,$$

where

$$A_1\bigg/\left(\frac{T_3}{|a_3|} - \frac{T_2}{a_2}\right) = A_2\bigg/\left(\frac{T_1}{a_1} - \frac{T_3}{|a_3|}\right) = A_3\bigg/\left(\frac{T_2}{a_2} - \frac{T_1}{a_1}\right),$$

and the *discrete* values of R for marginal stability satisfy the relation

$$\frac{a_1 T_1}{a_1^2 + \pi^2}\left(\frac{T_3}{|a_3|} - \frac{T_2}{a_2}\right) + \frac{a_2 T_2}{a_2^2 + \pi^2}\left(\frac{T_1}{a_1} - \frac{T_3}{|a_3|}\right)$$
$$+ \frac{|a_3| T_3}{a_3^2 - \pi^2}\left(\frac{T_2}{a_2} - \frac{T_1}{a_1}\right) = 0.$$

Here we have used the definitions

$$T_1 = \tan a_1 L, \qquad T_2 = \tan a_2 L, \qquad T_3 = \tanh |a_3| L.$$

[Drazin (1975).]

6.8 *Heat transfer in weakly nonlinear Rayleigh–Bénard convection.* Show that the Nusselt number

$$Nu = Hd/k(\theta_0 - \theta_1),$$

where H is the actual heat transfer into the fluid per unit area of plate, $k = c\rho\kappa$ is the thermal conductivity, and c is the specific heat of the fluid.

Using equation (6.22), show that

$$H = d^{-1}k(\theta_0 - \theta_1)(1 + bA),$$

when $R \approx R_c$, where b is some dimensionless constant that depends on the cell shape and A satisfies Landau equation (6.23). Evaluate the Nusselt number as a function of the Rayleigh number. Compare your result with the experimental measurements of Figure 6.4.

6.9 *The energy method for Rayleigh–Bénard convection.* You are given that the dimensionless nonlinear equations governing *perturbations* of the basic state of rest with heat conduction, namely $\mathbf{U} = \mathbf{0}$ and $\Theta = -z$, in Rayleigh–Bénard convection are

$$\frac{\partial \mathbf{u}}{\partial t} + \mathbf{u} \cdot \nabla \mathbf{u} = -\nabla p + R Pr \theta \mathbf{k} + Pr \Delta \mathbf{u},$$

$$\nabla \cdot \mathbf{u} = 0,$$

$$\frac{\partial \theta}{\partial t} - w + \mathbf{u} \cdot \nabla \theta = \Delta \theta.$$

Supposing that the flow satisfies free–free boundary conditions at $z = 0, 1$, and has period $2\pi/a_x$ in x and period $2\pi/a_y$ in y, deduce the 'energy' equation of the perturbation,

$$\frac{d(H + PrRK)}{dt} = 2Pr \int_{\mathcal{V}} \left[R\theta w - \frac{1}{2}\left(\frac{\partial u_i}{\partial x_j}\right)^2 - \frac{1}{2}R\left(\frac{\partial \theta}{\partial x_j}\right)^2 \right] dx,$$

(E6.4)

where

$$K = \int_{\mathcal{V}} \frac{1}{2} u_i^2 \, dx, \qquad H = \int_{\mathcal{V}} \frac{1}{2}\theta^2 \, dx,$$

and the domain of integration is the cuboid $\mathcal{V} = \{\mathbf{x} : 0 \le x < 2\pi/a_x, 0 \le y < 2\pi/a_y, 0 < z < 1\}$.

Using the constraint that $\partial u_i / \partial x_i = 0$ with the Lagrange multiplier p, show that the variational principle to minimize the integral on the right-hand side of the 'energy' equation (E6.4) gives the Euler–Lagrange equations

$$0 = -\nabla p + R\theta \mathbf{k} + \Delta \mathbf{u}$$

and

$$-w = \Delta\theta,$$

that is, the linearized equations (6.6), (6.8) for *steady* perturbations. Hence show that all perturbations, whether large or small, decay (that is to say, that the flow is globally asymptotically stable) if $R < R_c$, where R_c is the critical value of the Rayleigh number determined by the linear theory with free–free boundary conditions at $z = 0, 1$. [After Drazin & Reid (1981, Problem 7.13); Joseph (1965).]

6.10 *Derivation of the Lorenz system as a model of thermal convection.* You are *given* that the perturbations $\mathbf{u}$ and θ of the velocity and temperature fields respectively of a fluid heated from below are governed by the Boussinesq equations, namely,

$$\frac{\partial \mathbf{u}}{\partial t} + \mathbf{u} \cdot \nabla \mathbf{u} = -\nabla p + \sigma\theta\mathbf{k} + \sigma\Delta\mathbf{u},$$

$$\nabla \cdot \mathbf{u} = 0,$$

$$\frac{\partial \theta}{\partial t} + \mathbf{u} \cdot \nabla\theta = Rw + \Delta\theta,$$

for $-\infty < x < \infty, 0 \le z \le \pi$, where $\sigma = \nu/\kappa$ is the Prandtl number of the fluid, $R = \alpha g d^3 \Delta T/\kappa\nu$ is the Rayleigh number, and dimensionless variables are used with units d of length, d^2/κ of time and $\Delta T/R$ of temperature. Here πd is the depth of the layer of fluid, ΔT the temperature difference imposed across the layer, α the coefficient of thermal expansion, κ the thermal diffusivity and ν the kinematic viscosity of the fluid. There are 'free' perfectly conducting boundaries at $z = 0, \pi$, so

$$\frac{\partial u}{\partial z} = w = \theta = 0 \quad \text{at } z = 0, \pi.$$

You are further given that there are weakly nonlinear roll cells of the approximate form

$$u(x, z; t) = 2^{1/2}(k^2 + 1)k^{-1}X(t)S_xC_z,$$

$$w(x, z; t) = -2^{1/2}(k^2 + 1)X(t)C_xS_z,$$

$$\theta(x, z; t) = -(k^2 + 1)^3k^{-2}[2^{1/2}Y(t)C_xS_z + Z(t)S_{2z}],$$

where $S_x = \sin kx$, $C_z = \cos z$, $C_x = \cos kx$, $S_z = \sin z$ and $S_{2z} = \sin 2z$.

(i) Verify that the equation of continuity is satisfied.

(ii) Verify that the boundary conditions are satisfied.

(iii) Show that each nonlinear term in the momentum equation is proportional to $\sin 2kx$ or S_{2z}. Hence or otherwise show that the curl of the curl of the momentum equation gives

$$\frac{dX}{d\tau} = \sigma(Y - X), \tag{E6.5}$$

if appropriate components may be truncated, where $\tau = (k^2 + 1)t$. [Hint: Show that $\partial u/\partial z - \partial w/\partial x = -2^{1/2}(k^2 + 1)^2k^{-1}XS_xS_z$.]

(iv) Similarly, show that

$$\frac{dY}{d\tau} = rX - Y - ZX, \tag{E6.6}$$

$$\frac{dZ}{d\tau} = -bX + XY, \tag{E6.7}$$

where $r = k^2R/(k^2 + 1)^3$, $b = 4/(k^2 + 1)$. [Hint: Show that $\mathbf{u} \cdot \nabla\theta = (k^2 + 1)^4k^{-2}(XYS_{2z} + 2^{3/2}ZXC_xS_zC_{2z})$. Lorenz (1963).]

6.11 *Solutions of the Lorenz system.* Use the Lorenz system (E6.5)–(E6.7) of the previous question as a model problem of stability, bifurcation, symmetry breaking, onset of chaos and symmetry mending as follows.

Find when the null solution of the Lorenz system is stable, and explain the physical significance of your findings in terms of Rayleigh–Bénard convection.

Find all the other steady solutions of the Lorenz system. Discuss the physical significance of your findings in terms of Rayleigh–Bénard convection.

*Discuss the stability of these other steady solutions and the changes of the set of attractors as r increases from zero to infinity for fixed σ, b. [Hint: You are likely to need to read a book, for example, Drazin (1992, §8.1), to do this.]

It has been suggested that the Lorenz system models certain features of turbulent convection. Discuss its merits and demerits as a model.

6.12 *The Swift–Hohenberg model of nonlinear thermal convection.* You are *given* that the perturbations w of some velocity are governed by the equation

$$\frac{\partial w}{\partial t} = \left[R - R_c - \left(\frac{\partial^2}{\partial x^2} + 1 \right)^2 \right] w - w^3$$

for $-\infty < x < \infty$ and $t \geq 0$, where R is a parameter modelling the Rayleigh number. Find the linear stability characteristics of the null solution by the method of normal modes $\propto e^{ikx+st}$. Sketch the marginal curve in the (R, k)-plane. What is the critical value of R?

Find other solutions which are independent of x, t and identify a pitchfork bifurcation.

Assuming that w has period 2π in x, and expressing the solution of the equation as the complex Fourier series

$$w(x, t) = \sum_{n=-\infty}^{\infty} w_n(t) e^{inx},$$

show that $w_{-n} = w_n^*$. Find an infinite system of ordinary differential equations for $\{w_n\}$. Verify the stability of the null solution and find when the other solutions are stable.

Assuming that

$$w(x, t) = (R - R_c)^{1/2} [W_1(t) e^{ix} + W_{-1}(t) e^{-ix}]$$
$$+ O(R - R_c) \quad \text{as } R \to R_c+,$$

deduce plausibly the Landau equation for a supercritical bifurcation,

$$\frac{dW_1}{dT} = W_1 - |W_1|^2 W_1,$$

where $T = (R - R_c)t$. [Swift & Hohenberg (1977).]

6.13 *Rayleigh–Darcy convection in a porous medium.* You are *given* that two-dimensional convection in an infinite layer of a Boussinesq fluid in a porous medium is governed by the following dimensionless boundary-value problem:

$$\Delta \psi = -R \frac{\partial \theta}{\partial x} \quad \text{(this is the curl of Darcy's law)},$$

$$\frac{\partial \theta}{\partial t} + \frac{\partial \psi}{\partial z} \frac{\partial \theta}{\partial x} - \frac{\partial \psi}{\partial x} \frac{\partial \theta}{\partial z} = \Delta \theta \quad \text{(heat equation)},$$

$$\psi = 0, \quad \theta = -1 \quad \text{at } z = 1 \text{ (rigid perfectly conducting top)},$$

$$\psi = 0, \quad \theta = 0 \quad \text{at } z = 0 \text{ (rigid perfectly conducting bottom)},$$

where R is a modified Rayleigh number. Show that $\Psi = 0, \Theta = -z$ gives a basic state of rest, representing uniform conduction of heat from the bottom to the top of the layer.

Taking perturbations $\psi = \Psi + \psi', \theta = \Theta + \theta'$, linearize the system. Thence show that with normal modes of the form $\psi' = e^{st+ikx}\hat{\psi}(z)$, $\theta' = e^{st+ikx}\hat{\theta}(z)$, the eigenvalue relation is

$$s = \frac{k^2 R}{k^2 + n^2\pi^2} - \left(k^2 + n^2\pi^2\right) \quad \text{for } n = 1, 2, \ldots.$$

Deduce that the flow is unstable if $R > R_c = 4\pi^2$, the most unstable mode having wavelength 2 in the x-direction. [Horton & Rogers (1945), Lapwood (1948).]

6.14 *Weakly nonlinear Rayleigh–Darcy convection in a porous medium.* (i) Show that the equations of the previous Exercise 6.13 give, without further approximation,

$$\Delta \theta' - \frac{\partial \psi'}{\partial x} = \epsilon^2 \frac{\partial \theta'}{\partial T} + \frac{\partial \psi'}{\partial z} \frac{\partial \theta'}{\partial x} - \frac{\partial \psi'}{\partial x} \frac{\partial \theta'}{\partial z}, \quad \Delta \psi' + R_c \frac{\partial \theta'}{\partial x} = -\epsilon^2 \frac{\partial \theta'}{\partial x},$$

and

$$\psi' = \theta' = 0 \quad \text{at } z = 0, 1,$$

as the strongly nonlinear equations of the perturbation, where we define $\epsilon = (R - R_c)^{1/2}$ and $T = \epsilon^2 t$.

(ii) Assuming that θ', ψ' have period 2 in x (because, by the linear theory, the most unstable mode has wavenumber $k = \pi$); defining the amplitude

$$A(T) = 2\epsilon^{-1} \int_0^1 \int_0^2 \psi' \cos \pi x \sin \pi z \, dx \, dz$$

(to normalize the streamfunction); imposing (without loss of generality, because it is equivalent merely to a translation of x) the *phase condition*

$$\int_0^1 \int_0^2 \psi' \sin \pi x \sin \pi z \, dx \, dz = 0;$$

and assuming that

$$\theta' = \epsilon \theta_1 + \epsilon^2 \theta_2 + \cdots, \qquad \psi' = \epsilon \psi_1 + \epsilon^2 \psi_2 + \cdots \qquad \text{as } \epsilon \to 0;$$

show that

$$\Delta \theta_1 - \frac{\partial \psi_1}{\partial x} = 0, \quad \Delta \psi_1 + R_c \frac{\partial \theta_1}{\partial x} = 0 \qquad \text{for } 0 < x < 2, \, 0 < z < 1,$$

$$\psi_1 = \theta_1 = 0 \quad \text{at } z = 0, 1,$$

$$A(T) = 2 \int_0^1 \int_0^2 \psi_1 \cos \pi x \sin \pi z \, dx \, dz,$$

$$\int_0^1 \int_0^2 \psi_1 \sin \pi x \sin \pi z \, dx \, dz = 0.$$

Deduce that

$$\theta_1 = \tfrac{1}{2} \pi^{-1} A \sin \pi x \sin \pi z, \qquad \psi_1 = A \cos \pi x \sin \pi z,$$

and $R_c = 4\pi^2$.

(iii) Show that if

$$\Delta \theta_n - \frac{\partial \psi_n}{\partial x} = F_n, \quad \Delta \psi_n + R_c \frac{\partial \theta_n}{\partial x} = G_n \quad \text{for } 0 < x < 2, \, 0 < z < 1,$$

and

$$\psi_n = \theta_n = 0 \quad \text{at } z = 0, 1,$$

for some well-behaved functions $F_n(x, z)$, $G_n(x, z)$, then

$$\Delta^2 \psi_n + R_c \frac{\partial^2 \psi_n}{\partial x^2} = \Delta G_n - R_c \frac{\partial F_n}{\partial x}.$$

Integrating by parts repeatedly, deduce that

$$\int_0^1 \int_0^2 \left(\Delta G_n - R_c \frac{\partial F_n}{\partial x} \right) \psi_1 \, dx \, dz$$

$$= \int_0^1 \int_0^2 \psi_n \left(\Delta^2 \psi_1 - R_c \frac{\partial^2 \psi_1}{\partial x^2} \right) dx \, dz - \int_0^2 \left[G_n \frac{\partial \psi_1}{\partial z} \right]_{z=0}^1 dx.$$

Hence show that a solvability condition for the existence of θ_n, ψ_n is that

$$\int_0^1 \int_0^2 \left(\Delta G_n - R_c \frac{\partial F_n}{\partial x} \right) \psi_1 \, dx \, dz = - \int_0^2 \left[G_n \frac{\partial \psi_1}{\partial z} \right]_{z=0}^1 dx.$$

(iv) Show that

$$\theta_2 = -\tfrac{1}{16} \pi^{-1} A^2 \sin 2\pi z, \qquad \psi_2 = 0.$$

(v) Show that

$$\Delta \theta_3 - \frac{\partial \psi_3}{\partial x} = \frac{\partial \theta_1}{\partial T} + \frac{\partial \psi_1}{\partial z} \frac{\partial \theta_2}{\partial x} + \frac{\partial \psi_2}{\partial z} \frac{\partial \theta_1}{\partial x} - \frac{\partial \psi_1}{\partial x} \frac{\partial \theta_2}{\partial z} - \frac{\partial \psi_2}{\partial x} \frac{\partial \theta_1}{\partial z},$$

$$\Delta \psi_3 + R_c \frac{\partial \theta_3}{\partial x} = -\frac{\partial \theta_1}{\partial x} \quad \text{for } 0 < x < 2, \, 0 < z < 1,$$

and

$$\psi_3 = \theta_3 = 0 \quad \text{at } z = 0, 1;$$

and deduce the Landau equation,

$$\frac{dA}{dT} = \frac{1}{2} A - \frac{1}{8} \pi^2 A^3,$$

for a supercritical pitchfork bifurcation. [Palm *et al.* (1972).]

6.15 *A simple linear ordinary-differential system which models an aspect of double-diffusive convection.* Given that

$$\frac{dw}{dt} = -4w - k\theta - 4\epsilon w, \qquad \frac{d\theta}{dt} = 2\theta + kw - \epsilon\theta,$$

show that there are normal modes with $w, \theta \propto e^{st}$, where

$$s = -1 - \tfrac{5}{2}\epsilon \pm \tfrac{1}{2}\left[9(\epsilon + 2)^2 - 4k^2 \right]^{1/2}.$$

Deduce that for $k^2 > 8$ the null solution is stable if $\epsilon = 0$ and if $\epsilon \to \infty$ for fixed k, yet is unstable if $8 < k^2 < 9$ and $(2\epsilon - 1)^2 < 9 - k^2$.

[It can be seen that a state (for $8 < k^2 < 9$) is stable for $\epsilon = 0$ and for large ϵ, yet is unstable for some positive values of ϵ; thus a so-called 'stabilizing force' may destabilize a flow. Hinch (1973).]

Discuss this as a model of *double-diffusive convection*, the phenomenon when two quantities (perhaps heat and one solute, or two solutes) are diffused at different rates.

6.16 *An unbounded steady double-diffusive nonlinear wave.* You are given that the perturbations of a basic state of rest of a Boussinesq fluid with uniform gradients of temperature and of concentration of a solute are governed by the equations

$$\frac{\partial(\Delta\psi')}{\partial t} + J(\psi', \Delta\psi') = Pr\left[\frac{\partial(R\theta' - R_s c')}{\partial x} + \Delta^2\psi'\right],$$

$$\frac{\partial\theta'}{\partial t} + J(\psi', \theta') + \frac{\partial\psi'}{\partial x} = \Delta\theta',$$

$$\frac{\partial c'}{\partial t} + J(\psi', c') + \frac{\partial\psi'}{\partial x} = S\Delta c',$$

where $u' = \partial\psi'/\partial z$, $w' = -\partial\psi'/\partial x$, the Jacobian J is defined by $J(\psi, \phi) = (\partial\psi/\partial z)(\partial\phi/\partial x) - (\partial\psi/\partial x)(\partial\phi/\partial z)$ for all differentiable functions ψ, ϕ, the dimensionless Prandtl number $Pr = \nu/\kappa$, Schmidt number $S = \kappa_s/\kappa$, Rayleigh number $R = \alpha\beta g d^4/\kappa\nu$, and solute Rayleigh number $R_s = lg(\Delta c)d^3/\kappa\nu$, and κ_s is the diffusivity of the solute of concentration c.

Verify that an exact steady nonlinear solution is given by

$$\psi' = -\widehat{W}\cos x, \qquad \theta' = \hat{\theta}\sin x, \qquad c' = \hat{c}\sin x$$

for $-\infty < x, z < \infty$, provided that $R_s = S(R + 1)$, where the constants $\widehat{W}, \hat{\theta}, \hat{c}$ satisfy

$$\widehat{W} = R\hat{\theta} - R_s\hat{c}, \qquad \hat{\theta} = -\widehat{W}, \qquad \hat{c} = -\widehat{W}/S.$$

[Stern (1960).]

6.17 *Instability and convection due to variation of surface tension with temperature.* Consider the linear stability of a layer of uniform incompressible heat-conducting viscous fluid with negligible buoyancy confined by a rigid plate at $z_* = 0$ and a free surface at $z_* = d$.

Suppose that the plate is a perfect conductor of heat, and is maintained at a constant uniform temperature θ_0, and the free surface is a perfect insulator with basic temperature θ_1, and with surface tension $\gamma = \gamma_0 - c\theta_*$ for some constant $c > 0$.

Show that the conditions of tangential stress at the surface give

$$\mu \frac{\partial u'_{*i}}{\partial z_*} = \frac{\partial \gamma}{\partial x_{*i}} = -c \frac{\partial \theta_*}{\partial x_{*i}} \quad \text{at } z_* = d \text{ for } i = 1, 2;$$

but that the equations and other boundary conditions are the same as in §6.2 for $R = 0$ and free–rigid boundaries.

Taking normal modes of wavenumber a and assuming the principle of exchange of stabilities, show that at marginal stability

$$W \propto \sinh az + (a \coth a - 1)z \sinh az - az \cosh az$$

and

$$M = 8a^2(\cosh a \sinh a - a) \cosh a / (\sinh^3 a - a^3 \cosh a),$$
$$= f(a),$$

say, where M is the *Marangoni number* defined as $M = \beta c d^2 / \kappa \mu$. Deduce that there is stability if and only if $M \leq \min F(a) = 79.607$, the minimum being attained when $a = 1.993$. [After Drazin & Reid (1981, Problem 2.13); Pearson (1958).]

6.18 *A homely experiment on thermal instability.* Pour a light oil (corn oil serves well) in a clean deep frying pan (i.e. a skillet) so that there is a layer of oil about 2 mm deep. Heat the bottom of the pan gently and uniformly. To visualize the instability, drop in a little powder gently (cocoa serves well). Sprinkling powder on the surface reveals the pattern of steady polygonal cells. The motion of individual particles of the powder may be seen, with rising near the centre of a cell and falling near the sides. Tilt the pan a little to reveal the critical depth for the occurrence of convection. Verify that the size of the cells and the critical Rayleigh number are of the order of magnitude predicted by Rayleigh's theory. [After Drazin & Reid (1981, Problem 2.5).]

6.19 *A simple experiment on thermohaline convection.* Half fill a beaker with hot dyed brine (a few grams of salt and a very little ink or methylene blue in a half a litre of water in a litre beaker would serve well). Insert cold fresh water into the brine. To restrict the mixing of the water and the brine while you do this, pour the water slowly through a tube close to the bottom of the beaker, for example as shown in Figure 6.8. Rest the beaker on white paper (to see the ensuing convection clearly), keep it steady

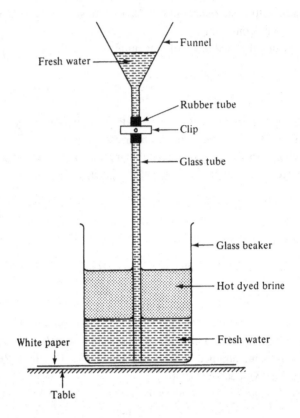

Figure 6.8   Sketch of the apparatus for a simple experiment on thermohaline convection. See Exercise 6.19. (After Drazin & Reid, 1981, Fig. 2.7.)

(to avoid setting up the motion of internal gravity waves) and ensure that there is no swirl (to avoid twisting the convection cells). The density of the hot brine should be less than that of the cold water, so that the liquids are in hydrostatically stable equilibrium. However, *thermohaline instability* will soon arise owing to the different rates of diffusion of heat and salt, the diffusion of salt being very slow. Observe the finger-like cells grow at a rate of order of magnitude 10^{-1} mm s^{-1} with a horizontal scale of about 3 mm. Discuss this as an example of double-diffusive convection, the more general phenomenon when two quantities (perhaps heat and one solute, as here, or two solutes) are diffused at different rates. [After Drazin & Reid (1981, Problem 2.6); Stern (1960), Baines & Gill (1969).]

7

Centrifugal Instability

The wind goeth toward the south, and turneth about unto the north;
it whirleth about continually: and the wind returneth according to
his circuits.

Eccl. i 6

Centrifugal force, an analogue of buoyancy as a force field which may create
or suppress instability of flows, is examined in this chapter just as buoyancy
was in the last chapter. The prototypical problem of instability of Couette flow
between two rotating coaxial cylinders is described, but not in full detail.

7.1 Swirling Flows

Another important type of hydrodynamic instability is that of axisymmetric
swirling flows. They are analogous to thermal convection, the centrifugal force
being the analogue of the buoyancy. By such swirling flows, we mean those
with basic velocity components and pressure of the form $\mathbf{u} = \mathbf{U}(r)$, $p = P(r)$,
where

$$U_r = 0, U_\theta = V(r), U_z = 0 \quad \text{for} \quad -\infty < z < \infty, \tag{7.1}$$

in cylindrical polar coordinates (r, θ, z). (See Figure 7.1 for the configuration
and notation.) This gives an exact solution of the Euler equations of motion of
an incompressible inviscid fluid for all functions V. But for the Navier–Stokes
equations of a viscous fluid we require

$$V(r) = Ar + B/r. \tag{7.2}$$

For Couette flow, we take rigid cylinders $r = R_1, R_2$ with angular velocities
Ω_1, Ω_2 respectively, and deduce that

$$A = \frac{\Omega_2 R_2^2 - \Omega_1 R_1^2}{R_2^2 - R_1^2}, \qquad B = \frac{\Omega_1 - \Omega_2}{R_1^{-2} - R_2^{-2}}. \tag{7.3}$$

However, let us first consider a physical argument of Rayleigh (1916b) for
an inviscid fluid. The angular momentum per unit mass of a ring element
of fluid $r = $ constant, $z = $ constant in the basic swirling flow is $H = rV(r)$,

123

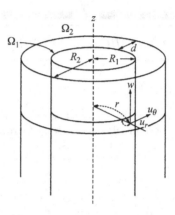

Figure 7.1   Sketch of the cross-section of the long coaxial cylinders of Couette flow, with notation. (After P. Drazin and T. Kambe, *Ryutai Rikigaku – Anteisei To Ranyu (Fluid Dynamics – Stability and Turbulence)*, University of Tokyo Press, 1989, Fig. 6.1. Reproduced by permission of the University of Tokyo Press.)

and the circulation of the ring is $2\pi H$. But, by Kelvin's circulation theorem, the circulation will be conserved when the ring is perturbed, and so H will be. The swirl is manifest as a centrifugal force density $\rho V^2/r = \rho H^2/r^3$, which acts in the radial direction and which is associated with a potential energy density $\frac{1}{2}\rho H^2/r^2 = \frac{1}{2}\rho V^2$. With these facts in mind, consider the effects of an interchange of two elemental rings of equal mass, at $r = r_1, z = z_1$ and $r = r_2, z = z_2$, say. It follows that their kinetic energy per unit volume is $\frac{1}{2}\rho(H_1^2\bar{r}_1^2 + H_2^2\bar{r}_2^2)$ before and $\frac{1}{2}\rho(H_1^2 r_2^{-2} + H_2^2 r_1^{-2})$ after the interchange. The increase in kinetic energy, required to effect the interchange, is therefore proportional to

$$(H_2^2 - H_1^2)(r_1^{-2} - r_2^{-2}).$$

Take $r_2 > r_1$ without loss of generality. Then the interchange can only release energy, and hence generate instability, if $H_1^2 > H_2^2$. Thus there may be instability only if $H_1^2 > H_2^2$ for some $r_1 < r_2$, that is, only if

$$\frac{\mathrm{d}(r^2 V^2)}{\mathrm{d}r} < 0 \tag{7.4}$$

somewhere in the flow. It is called *Rayleigh's criterion* for instability. Conversely, if

$$\frac{\mathrm{d}(r^2 V^2)}{\mathrm{d}r} > 0 \tag{7.5}$$

everywhere in the flow, then the flow is stable to all axisymmetric perturbations. It can be shown (Synge, 1933) mathematically that this is a necessary and sufficient condition for stability of *swirling* flow of an incompressible *inviscid* fluid to *axisymmetric* perturbations. However, the flow *may* be unstable to non-axisymmetric perturbations (which are not treated in Rayleigh's argument) if $d(r^2V^2)/dr > 0$ everywhere – you can see that there would be Kelvin–Helmholtz instability (as described in §3.8) to short waves if V were discontinuous, whether Rayleigh's criterion were satisfied or not, and so instability may plausibly be inferred to occur if V were to change smoothly but very rapidly.

7.2 Instability of Couette Flow

Taylor (1923), in a paper outstanding in the history of fluid dynamics, analysed the linear stability of Couette flow between rotating coaxial cylinders and verified his theory with experiments. First, just look at the instability shown in Figure 7.2(a) to see it. Next we shall explain his theory and experiment.

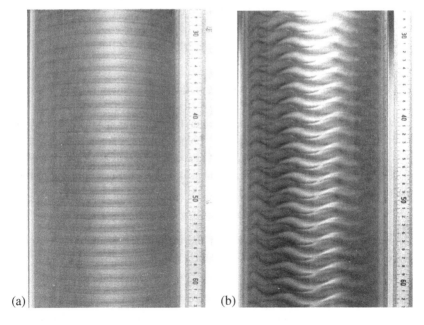

(a)　　　　　　　　　(b)

Figure 7.2　(a) A photograph of Taylor vortices, with the outer cylinder at rest and the inner cylinder rotating. (b) A photograph of wavy Taylor vortices, which occur at the onset of a secondary instability as the inner cylinder rotates more rapidly. (After Koschmieder, 1993, Figs. I.1, I.2.)

It is quite straightforward to

(1) take the basic Couette flow (7.2),
(2) linearize the Navier–Stokes equations and the boundary conditions for small perturbations of the basic flow,
(3) choose dimensionless variables and parameters,
(4) take normal modes of the form

$$\mathbf{u}'(\mathbf{x}, t) = \hat{\mathbf{u}}(r)e^{st+\mathrm{i}(n\theta+kz)}, \quad \text{and}$$

(5) thence derive an ordinary-differential eigenvalue problem to find s, $\hat{\mathbf{u}}$ for given real wavenumber k and integral wavenumber n.

The resulting numerical problem was too difficult in Taylor's time to solve. So he made some simplifying assumptions based on his experimental observations:

(1) He *assumed* that the most unstable perturbations are axisymmetric, and so put $n = 0$.
(2) He assumed that the principle of the exchange of stabilities is valid, that is, $\operatorname{Im}(s) = 0$ at the onset of instability, and so put $s = 0$ and sought dimensionless parameters which give the margin of stability.
(3) He assumed that there is a *narrow gap* between the cylinders, that is, $R_2 - R_1 \ll R_2$.

These three assumptions enabled him to reduce the solution of the governing eigenvalue problem to explicit terms of Bessel functions, and calculate the stability characteristics numerically. Later theoreticians have calculated the stability characteristics of the general problem, finding s for $n \neq 0$ as well as $n = 0$, for large ranges of R_1, R_2, Ω_1, Ω_2 and ν as well as k.

It is, of course, convenient to use dimensionless variables to present results. We define

$$\mu = \Omega_2/\Omega_1 \quad \text{and} \quad \eta = R_1/R_2. \tag{7.6}$$

To measure the *square* of the ratio of the magnitudes of inertia to viscous forces, we shall use what is now called the *Taylor number*,

$$T = -\frac{4ABR_1^2}{\nu^2} = \frac{4\Omega_1^2 R_1^4}{\nu^2} \frac{\left(1 - \Omega_2/\Omega_1\right)\left(1 - \Omega_2 R_2^2/\Omega_1 R_1^2\right)}{\left(1 - R_1^2/R_2^2\right)^2}; \tag{7.7}$$

you can see that it is essentially the square of a Reynolds number based on a circumferential velocity. Sometimes Reynolds numbers based on the circumferential velocities of the two cylinders are used instead. Rayleigh's swirl criterion suggests stability to axisymmetric perturbations in the limit as $v \rightarrow 0$ if and only if

$$0 < \frac{\mathrm{d}(rV)^2}{\mathrm{d}r} = \frac{\mathrm{d}\left(Ar^2 + B\right)^2}{\mathrm{d}r}$$
$$= 4Ar\left(Ar^2 + B\right),$$

that is, if and only if

$$\left(\Omega_2 R_2^2 - \Omega_1 R_1^2\right)V(r) > 0 \quad \text{for } R_1 < r < R_2,$$

that is, if and only if the outer cylinder has greater circulation than the inner cylinder *and* the cylinders rotate in the same direction. So we anticipate instability for a viscous fluid if the inner cylinder has greater circulation than the outer, or the cylinders rotate in opposite directions and the viscosity is not too large, that is, if T is greater than some critical value.

The marginal curve $s = 0$ in the (a, T)-plane, for fixed values of the other dimensionless parameters and for axisymmetric perturbations ($n = 0$), where the dimensionless wavenumber is $a = (R_2 - R_1)k$, has a similar shape to that of the curve of Figure 6.3 for Rayleigh–Bénard convection. But the other parameters are important here, so we illustrate the details of the stability characteristics, found by numerical calculations and confirmed by observations, best by Figures 7.3–7.6 below.

These linear stability characteristics suggest the onset of steady (because $\text{Im}(s) = 0$ when $\text{Re}(s) = 0$) axisymmetric (because the linear perturbations are usually most unstable when $n = 0$) toroidal vortices of axial wavelength $2\pi(R_2 - R_1)/a_c$ as T increases through T_c. This is what happens! See the photographs of Figure 7.2(a) and Van Dyke (1982, Figs. 127–128), and the two superb film loops of Coles (FL1963a,b). Look up the *Video Library* of Homsy *et al.* (CD2000), view the videos under the subheadings 'Steady Taylor Cells' and 'Taylor Instability'. Taylor's (1923) calculations and observations agreed with astonishing closeness of from 1 to 5 per cent, as shown in Figure 7.5. He observed steady toroidal vortices develop at the onset of instability, as the eigenfunctions showed him theoretically – these are now called *Taylor vortices*. (The flow is more complicated if Ω_2 is of different sign to Ω_1.)

Since 1923 there have been hundreds, perhaps thousands, of theoretical and experimental papers exploring scores of aspects of transition to turbulence in Couette flow. In particular, the weakly nonlinear theory (Davey, 1962) gives a pitchfork with a positive Landau constant to describe the primary instability which leads to the onset of supercritical Taylor vortices when the outer cylinder is at rest. As the Taylor number increases, the vortices themselves

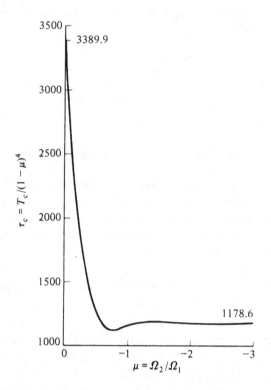

Figure 7.3   The critical Taylor number as a function of $\mu = \Omega_2/\Omega_1$ in the narrow-gap limit as $R_2/R_1 \to 1$. (After Harris & Reid, 1964, Fig. 2, and Drazin & Reid, 1981, Fig. 3.8.)

become unstable, and azimuthal (non-axisymmetric) waves develop on them as a secondary instability (see a picture of wavy Taylor vortices in Figure 7.2(b)). See also the photographs of Van Dyke (1982, Figs. 127, 128, 130), and relate them to the results of the theory. Do the same with the measurements of the torque as a function of the Taylor number shown in Figure 7.7.

As the Taylor number increases further, a succession of bifurcations with chaos occur in a complicated transition to turbulence (see Fenstermacher *et al.*, 1979). It is especially interesting that sometimes chaos may ensue, abate and ensue again in the succession of bifurcations as the Taylor number increases. The route to turbulence depends on R_2/R_1, Ω_2/Ω_1 *and* the length of the cylinders. Some of this can be seen in the various regimes of flow observed for various values of the Reynolds numbers, $Re_1 = R_1^2\Omega_1/\nu$ and $Re_2 = R_2^2\Omega_2/\nu$ for two longish cylinders, shown in Figure 7.8. To find more of the experimental

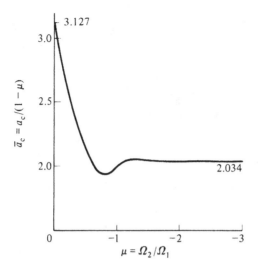

Figure 7.4 The critical dimensionless wavenumber $\bar{a}_c = a_c/(1 - \mu)$ as a function of $\mu = \Omega_2/\Omega_1$ in the narrow-gap limit as $R_2/R_1 \to 1$. (After Harris & Reid, 1964, Fig. 1, and Drazin & Reid, 1981, Fig. 3.9.)

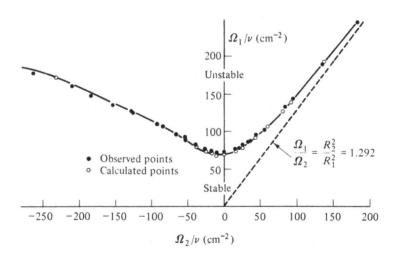

Figure 7.5 Sketch (after Taylor, 1923) of the marginal curve in the $(\Omega_1/\nu, \Omega_2/\nu)$-plane for *his* pair of cylinders, i.e. for $R_1 = 3.55$ cm and $R_2 = 4.035$ cm. (After Drazin & Reid, 1981, Fig. 3.13.)

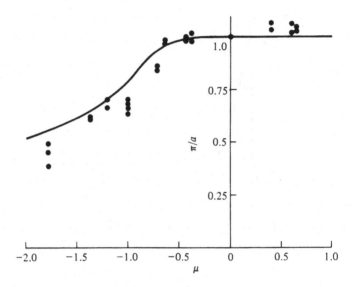

Figure 7.6   Observations and narrow-gap calculations of Taylor vortices in water for cylinders with $R_1 = 3.80$ cm and $R_2 = 4.035$ cm. Spacing of Taylor vortices at the onset of instability. (After Drazin & Reid, 1981, Fig. 3.12; experimental data are from Taylor, 1923, Table 8.)

results and their relationship to the theoretical results of both the linear and weakly nonlinear theories, read Koschmieder's (1993) monograph.

The bifurcations for short cylinders, in distinction to the long ones of Taylor's idealized model, also have been studied, notably by Benjamin and Mullin (1982). The critical values of the Taylor number at the bifurcations are more widely separated than for long cylinders, and some bifurcations are more easy to observe in experiments. It is especially interesting that Benjamin & Mullin (1982) observed as many as 20 coexisting different stable steady flows (with different numbers or senses of rotation of vortices between the two ends of the pair of cylinders), and inferred the existence of 19 more unstable steady flows, all for the same steady experimental configuration. Again, Koschmieder (1993) describes the details.

7.3 Görtler Instability

In addition to Taylor vortices, there is another important type of centrifugal instability, which was discovered by Görtler in 1940. Görtler treated the instability of a boundary layer on a concave wall, as illustrated in Figure 7.9. The instability, now called *Görtler instability*, is manifest as *Görtler vortices*. He made three approximations in the modelling. The thickness, say δ, of the

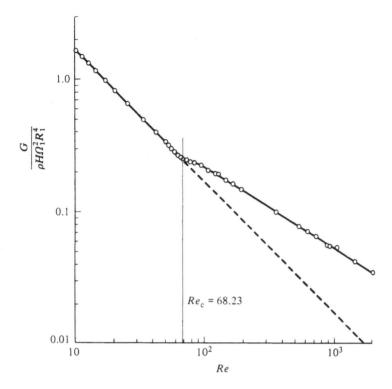

Figure 7.7   Variation of the torque G with the Reynolds number $Re = \Omega_1 R_1 (R_2 - R_1)/\nu$ for $\Omega_2 = 0$, $R_1 = 1.0$ cm, $R_2 = 2.0$ cm, $H = 5.0$ cm, $\nu = 0.1226$ cm^2 s^{-1}, $\rho = 0.8404$ g cm^{-3}. The torque G is measured in dyne cm ($=10^{-7}$ N m). (After Drazin & Reid, 1981, Fig. 3.14, using data of Donnelly & Simon, 1960, Table 2.)

boundary layer is assumed to be much smaller than the radius, say R_0, of curvature of the wall. Also the basic flow is nearly parallel to the wall, so that the centrifugal force is neglected in determining the basic flow, although retained in the linearized equations for the perturbation. Further, the basic flow is assumed to depend only on the coordinate z transverse to the wall by a quasi-parallel approximation.

We choose δ as the length scale, the velocity U_∞ of a given uniform outer stream as the velocity scale, and $\eta = y/\delta$ as the coordinate normal to the wall to make the problem dimensionless. Then we take a dimensionless basic velocity

$$\mathbf{U} = U(\eta)\mathbf{i} \quad \text{for } 0 \leq \eta,$$

where $U(\eta) \to 1$ as $\eta \to \infty$.

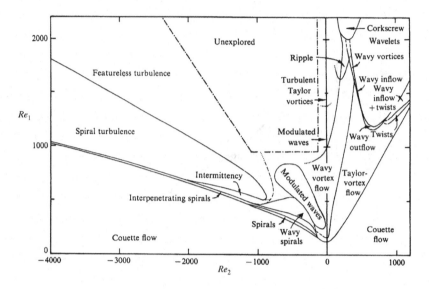

Figure 7.8   Regimes of flow observed for two long rotating coaxial rigid cylinders with Reynolds numbers $Re_1 = R_1^2 \Omega_1 / \nu$ and $Re_2 = R_2^2 \Omega_2 / \nu$. (After Andereck *et al.*, 1986, Fig. 1.)

Linearizing the Navier–Stokes equations for the perturbed flow $\mathbf{u} = \mathbf{U} + \mathbf{u}'$, $p = P + p'$, taking a normal mode of the form

$$(u', v', w', p') = (u(\eta), v(\eta), w(\eta), p(\eta))e^{st+ikx}, \qquad (7.8)$$

and substituting into the linearized equations, we find four coupled ordinary differential equations. On eliminating w, p from these, we derive the system

$$\left(D^2 - a^2\right)\left(D^2 - a^2 - \sigma\right)v = -a^2 \mu U u, \qquad (7.9)$$

$$\left(D^2 - a^2 - \sigma\right)u = U'v, \qquad (7.10)$$

where

$$D = d/d\eta, \quad a = k\delta, \quad \sigma = s\delta^2/\nu, \quad \mu = 2(U_\infty \delta/\nu)^2, \quad U' = DU. \qquad (7.11)$$

The boundary conditions of impermeability and no slip give

$$u(\eta) = v(\eta) = Dv(\eta) = 0 \quad \text{at } \eta = 0 \text{ and as } \eta \to \infty. \qquad (7.12)$$

The dimensionless parameter μ is an analogue of the narrow-gap Taylor number. However, it is customary to express the results instead in terms of the

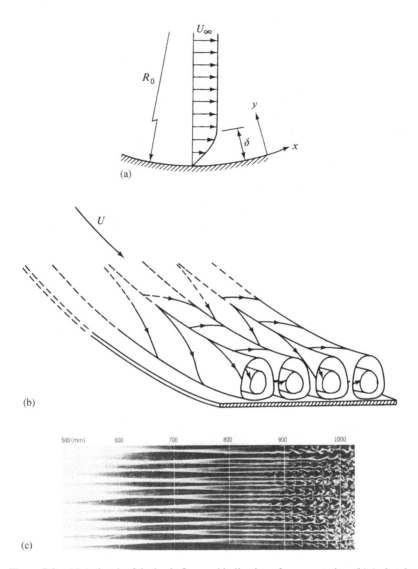

Figure 7.9 (a) A sketch of the basic flow and indication of some notation. (b) A sketch of the secondary flow which arises at the onset of instability in a boundary layer along a concave wall. (After Görtler, 1940, and Drazin & Reid, 1981, Fig. 3.20). (c) A photograph of Görtler instability. (After Nakayma, 1988, Fig. 30; reproduced by permission of Professor Akira Ito.)

Görtler number,

$$G = (U_\infty \vartheta / \nu)(\vartheta / R_0)^{1/2}, \qquad (7.13)$$

where ϑ is the momentum thickness of the boundary layer. It can be seen that $G^2 = \frac{1}{2}\theta^3 \mu / (\delta^2 R_0)$.

It is believed that the principle of exchange of stabilities is valid for this eigenvalue problem, that is, σ is real for marginally stable modes. So to determine marginal stability we put $\sigma = 0$ and the equations become

$$\left(D^2 - a^2\right)\left(D^2 - a^2\right)v = -a^2 \mu U u, \qquad \left(D^2 - a^2\right)u = U'v. \qquad (7.14)$$

For 40 years after the publication of Görtler's paper, there was much work on this eigenvalue problem, with calculations for various velocity profiles U. However, Hall (1983) showed that the assumption of a locally parallel flow is invalid unless the fluid is inviscid ($G = 0$) or the waves are short ($a \gg 1$), because the full linearized partial differential system has to be retained in all but the short-wave limit in order to take proper account of the spatial development of the boundary layer. It follows that a marginal curve or a growth rate at a given station has no meaning, but a parabolic partial differential system has to be solved as a marching problem. Saric (1994) has reviewed the experimental results and their relationship to the theory.

Exercises

7.1 *Stability problem for basic swirling flow of an incompressible inviscid fluid.* Supposing that $\mathbf{u} = \mathbf{U} + \mathbf{u}'$, $p = P + p'$, where $\mathbf{U} = (0, V(r), 0)$, $P = P(r)$, $\mathbf{u}' = (u'_r, u'_\theta, u'_z)$ in cylindrical polar coordinates (r, θ, z), and linearizing the Euler equations which govern the motion of an incompressible inviscid fluid, show that

$$\frac{\partial u'_r}{\partial t} + \Omega \frac{\partial u'_r}{\partial \theta} - 2\Omega u'_\theta = -\frac{1}{\rho} \frac{\partial p'}{\partial r},$$

$$\frac{\partial u'_\theta}{\partial t} + \Omega \frac{\partial u'_\theta}{\partial \theta} + \left(\frac{dV}{dr} + \frac{V}{r}\right)u'_r = -\frac{1}{\rho} \frac{1}{r} \frac{\partial p'}{\partial \theta},$$

$$\frac{\partial u'_z}{\partial t} + \Omega \frac{\partial u'_z}{\partial \theta} = -\frac{1}{\rho} \frac{\partial p'}{\partial z},$$

$$\frac{\partial u'_r}{\partial r} + \frac{u'_r}{r} + \frac{1}{r} \frac{\partial u'_\theta}{\partial \theta} + \frac{\partial u'_z}{\partial z} = 0,$$

where the local angular velocity of the basic flow is defined as $\Omega = V/r$.

Show that if the flow is bounded by rigid cylinders at $r = R_1$, R_2, then

$$u'_r = 0 \quad \text{at } r = R_1, R_2.$$

7.2 *Axisymmetric perturbations of a swirling flow.* Show that if $\partial/\partial\theta = 0$, then the linearized equations of Exercise 7.1 give

$$\frac{\partial^2}{\partial t^2} \left(\frac{\partial^2 u'_r}{\partial r^2} + \frac{1}{r}\frac{\partial u'_r}{\partial r} - \frac{u'_r}{r^2} + \frac{\partial^2 u'_r}{\partial z^2} \right) + \Phi\frac{\partial^2 u'_r}{\partial z^2} = 0,$$

where the *Rayleigh discriminant* of the basic flow is defined as $\Phi = r^{-3}\mathrm{d}(rV)^2/\mathrm{d}r$. [Thus Φ is an analogue of N^2, the square of the buoyancy frequency defined in Exercise 8.21.]

Taking a normal mode $(u'_r, u'_\theta, u'_z, p') = (u(r), 0, w(r), \varpi(r))\exp(st + ikz)$, show further that

$$(\mathrm{DD}_* - k^2)u - \frac{k^2}{s^2}\Phi u = 0,$$

where $\mathrm{D} = \mathrm{d}/\mathrm{d}r$, $\mathrm{D}_* = \mathrm{d}/\mathrm{d}r + 1/r$.
Show also that

$$u = 0 \quad \text{at } r = R_1, R_2.$$

By use of Sturm–Liouville theory, show that the eigenvalues k^2/s^2 are all negative if $\Phi > 0$ throughout the interval $R_1 < r < R_2$ and they are all positive if $\Phi < 0$ throughout the interval; however, if Φ changes sign, then there are both positive and negative eigenvalues k^2/s^2 with limit points at $\pm\infty$. [Synge (1933).]

Deduce that if $\Phi < 0$ somewhere, then the flow is unstable, in accordance with Rayleigh's swirl criterion.

7.3 *Two-dimensional perturbations of a swirling flow.* Again following on from Exercise 7.1, show that if u_z, $\partial/\partial z = 0$, $u'_r = \partial\psi'/r\partial\theta$, $u'_\theta = -\partial\psi'/\partial r$, then the linearized equations give

$$\left(\frac{\partial}{\partial t} + \Omega\frac{\partial}{\partial\theta} \right)\Delta\psi' - \frac{\mathrm{D}Z}{r}\frac{\partial\psi'}{\partial\theta} = 0,$$

where the basic vorticity $Z = r^{-1}\mathrm{D}(rV) = \mathrm{D}_*V$ and the Laplacian is given by

$$\Delta = \frac{\partial^2}{\partial r^2} + \frac{1}{r}\frac{\partial}{\partial r} + \frac{1}{r^2}\frac{\partial^2}{\partial\theta^2}.$$

[It is perhaps easier to derive this equation by linearizing the vorticity equation (2.4) with $R = \infty$.]

Taking normal modes with $\psi'(r, \theta, t) = \phi(r) \exp(st + in\theta)$, show that

$$(s + in\Omega)(D_*D - n^2/r^2)\phi - inr^{-1}(DZ)\phi = 0,$$

$$\phi = 0 \quad \text{at } r = R_1, R_2.$$

Deduce that a necessary condition for instability to two-dimensional perturbations is that the basic vorticity gradient DZ changes sign somewhere in the interval of flow.

[Rayleigh (1880). Hint: Adapt the proof (in §8.2) for the analogous problem of the instability of parallel flow.]

7.4 *The narrow-gap approximation.* Consider Couette flow with basic velocity $V(r) = Ar + B/r$, in the usual notation, and define the basic angular velocity $\Omega = V/r$. Deduce that

$$\Omega(r) \to \Omega_1[1 - (1 - \mu)\zeta] \quad \text{as } \eta \to 1$$

for fixed $\zeta = (r - R_1)/(R_2 - R_1)$, where $\eta = R_1/R_2$, $\mu = \Omega_2/\Omega_1$.

Deduce that the Rayleigh discriminant

$$\Phi(r) \sim -2\Omega_1^2 \frac{1 - \mu}{1 - \eta}[1 - (1 - \mu)\zeta] \quad \text{as } \eta \to 1.$$

7.5 *Instability of Couette flow of an inviscid fluid.* Show that if $\Omega(r) = \Omega_1[1 - (1 - \mu)\zeta]$ as in Exercise 7.4, then the stability of axisymmetric perturbations (Exercise 7.2) is governed by the eigenvalue problem

$$(D^2 - a^2)u = -a^2\sigma^{-2}[1 - (1 - \mu)\zeta]u,$$

$$u = 0 \quad \text{at } \zeta = 0, 1,$$

where $\sigma = s/(-4A\Omega_1)^{1/2} \sim s[2(1 - \eta)/(1 - \mu)]^{1/2}/2\Omega_1$, $a = k(R_2 - R_1)$.

Deduce that

$$\frac{d^2u}{dx^2} = xu, \quad u = 0 \quad \text{at } x = x_0, x_1,$$

where $x = [a\sigma^2/(1 - \mu)]^{2/3}\{1 - [1 - (1 - \mu)\zeta]/\sigma^2\}$, $x_0 = [a\sigma^2/(1 - \mu)]^{2/3}(1 - \sigma^{-2})$, $x_1 = [a\sigma^2/(1 - \mu)]^{2/3}(1 - \mu\sigma^{-2})$, and that the eigenvalue relation is

$$\frac{\text{Ai}(x_1)}{\text{Bi}(x_1)} = \frac{\text{Ai}(x_0)}{\text{Bi}(x_0)},$$

where Ai, Bi are the Airy functions.

[Numerical calculations confirm stability for $\mu > 1$ and instability for $\mu < 1$, in accordance with Rayleigh's criterion. Reid (1960); see Drazin & Reid (1981, §16).]

8

Stability of Parallel Flows

For they have sown the wind, and they shall reap the whirlwind. . . .

Hosea viii 7

This chapter is an account of the instability of many important flows: channel flows such as plane Couette flow and plane Poiseuille flow, Blasius's boundary layer on a flat plate, two-dimensional jets, wakes and free shear layers, and pipe flows and axisymmetric jets. It begins with the instability of plane parallel flows of an incompressible inviscid fluid, and goes on to the instability of those of a viscous fluid. We will see that for some flows the use of an inviscid fluid gives a good approximation to the stability characteristics of a viscous fluid at large values of the Reynolds number, but that for other flows it does not. Indeed, we will see that viscosity, although it dissipates energy, *may* destabilize a flow which is stable for an inviscid fluid. Let us begin then with the relatively simple theory for an incompressible inviscid fluid.

Part 1: Inviscid Fluid

8.1 Stability of Plane Parallel Flows of an Inviscid Fluid

First take scales V of velocity and L of length of the given basic plane parallel flow. For example, for a channel flow or a jet V might be the value of the velocity at the centre of the flow, and for a free shear layer or a boundary layer on a plate V might be the free stream velocity; and L might be the half width of a channel in which the fluid flows, or the width of a jet, free shear layer or boundary layer on a plate. Then define the dimensionless variables $\mathbf{x} = \mathbf{x}_*/L, t = Vt_*/L, \mathbf{u} = \mathbf{u}_*/V, p = p_*/\rho V^2$. Now Euler's equations of motion of an inviscid fluid may be written in dimensionless form as

$$\frac{\partial \mathbf{u}}{\partial t} + \mathbf{u} \cdot \nabla \mathbf{u} = -\nabla p, \qquad (8.1)$$

the equation of continuity for the incompressible fluid as

$$\nabla \cdot \mathbf{u} = 0, \qquad (8.2)$$

138

and boundary conditions of no velocity of penetration at rigid walls at $z = z_1, z_2$ as

$$w = 0 \quad \text{at } z = z_1, z_2. \tag{8.3}$$

We may take either $z_1 = -\infty$ or $z_2 = \infty$ if the flow is semibounded or both if the flow is unbounded.

Next take

$$\mathbf{U} = U(z)\mathbf{i}, \qquad P = \text{constant} \quad \text{at } z_1 \leq z \leq z_2, \tag{8.4}$$

to represent a basic flow along the channel (see Figure 8.1). It can be verified that this flow satisfies the governing equations and boundary conditions for all functions U.

To find the linear stability characteristics, linearize the equations and boundary conditions for small perturbations, and use the method of normal modes as usual. So express

$$\mathbf{u}(\mathbf{x}, t) = \mathbf{U}(z) + \mathbf{u}'(\mathbf{x}, t), \qquad p(\mathbf{x}, t) = P + p'(\mathbf{x}, t), \tag{8.5}$$

and linearize equations (8.1)–(8.3) by neglecting products of the small perturbed quantities (denoted by primes). It follows that

$$\left(\frac{\partial}{\partial t} + U \frac{\partial}{\partial x} \right) \mathbf{u}' + \frac{dU}{dz} w'\mathbf{i} = -\nabla p', \qquad \nabla \cdot \mathbf{u}' = 0, \tag{8.6}$$

$$w'(\mathbf{x}, t) = 0 \quad \text{at } z = z_1, z_2. \tag{8.7}$$

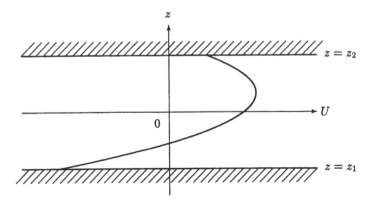

Figure 8.1   The configuration of the basic flow.

The divergence of equations (8.6) gives

$$\Delta p' = -2\frac{dU}{dz}\frac{\partial w'}{\partial x}.$$ (8.8)

It can thence be shown that the Laplacian of the z-component of equations (8.6) gives

$$\left(\frac{\partial}{\partial t} + U\frac{\partial}{\partial x}\right)\Delta w' - \frac{d^2 U}{dz^2}\frac{\partial w'}{\partial x} = 0.$$ (8.9)

All these equations have coefficients independent of x, y, t, but not of z, so we may separate the variables by taking independent normal modes of the form

$$\mathbf{u}'(\mathbf{x}, t) = \hat{\mathbf{u}}(z)e^{i(\alpha x + \beta y - \alpha ct)}, \qquad p'(\mathbf{x}, t) = \hat{p}(z)e^{i(\alpha x + \beta y - \alpha ct)},$$ (8.10)

it being understood as usual that each physical quantity is represented by the real part of its complex expression, so that, say, w' represents the physical quantity $[\text{Re}(\hat{w})\cos(\alpha x + \beta y - \alpha c_r t) - \text{Im}(\hat{w})\sin(\alpha x + \beta y - \alpha c_r t)]\exp(\alpha c_i t)$, where $c_r = \text{Re}(c)$, $c_i = \text{Im}(c)$. This mode is a wave travelling with phase velocity $\alpha c_r/(\alpha^2 + \beta^2)^{1/2}$ in the $(\alpha, \beta, 0)$-direction, while it decays or grows with time like $\exp(\alpha c_i t)$. The mode is accordingly said to be linearly stable if $\alpha c_i \leq 0$ and unstable if $\alpha c_i > 0$. (We could have written $-i\alpha c = s$ so that $\mathbf{u}'$, $p' \propto e^{st}$, but it is conventional to use the complex velocity c in this class of problems, for which travelling wave modes usually occur.)

Now equations (8.6) give

$$i\alpha(U - c)\hat{u} + \frac{dU}{dz}\hat{w} + i\alpha\hat{p} = 0,$$ (8.11)

$$i\alpha(U - c)\hat{v} + i\beta\hat{p} = 0,$$ (8.12)

$$i\alpha(U - c)\hat{w} + \frac{d\hat{p}}{dz} = 0,$$ (8.13)

$$i\alpha\hat{u} + i\beta\hat{v} + \frac{d\hat{w}}{dz} = 0.$$ (8.14)

Also boundary conditions (8.7) give

$$\hat{w}(\mathbf{x}, t) = 0 \quad \text{at } z = z_1, z_2.$$ (8.15)

Equations (8.11)–(8.14) and boundary conditions (8.15) pose an eigenvalue problem to determine all the eigenfunctions $\hat{\mathbf{u}}$, $\hat{p}$ belonging to eigenvalues c for given $U, z_1, z_2, \alpha, \beta$. The *flow* is linearly stable if $\alpha c_i \leq 0$ for *all* real wavenumbers α, β. Conversely, the flow is unstable if $\alpha c_i > 0$ for at least one pair of real wavenumbers α, β. The idea here is that any initial perturbation may be expressed as a superposition of eigenfunctions for all values of the wavenumbers, at least if the set of eigenfunctions is complete, and that the flow is stable if no mode grows without bound.

To solve the eigenvalue problem (8.11)–(8.15), it helps first to define

$$\tilde{\alpha} = \left(\alpha^2 + \beta^2\right)^{1/2}, \qquad \tilde{u} = (\alpha\hat{u} + \beta\hat{v})/\tilde{\alpha}, \qquad \tilde{p} = \tilde{\alpha}\hat{p}/\alpha. \qquad (8.16)$$

Equations (8.16) are the inviscid form of the *Squire's transformation*, which Squire (1933) originally used for a viscous fluid. Now take the sum of the products of α and equation (8.11) and of β and (8.12), and divide the sum by α to get

$$i\tilde{\alpha}(U - c)\tilde{u} + \frac{\mathrm{d}U}{\mathrm{d}z}\hat{w} + i\tilde{\alpha}\tilde{p} = 0. \qquad (8.17)$$

This is essentially the component of the linearized Euler equations in the direction of the wavenumber vector $\boldsymbol{\alpha} = \alpha\mathbf{i} + \beta\mathbf{j}$, for $\tilde{\alpha} = |\boldsymbol{\alpha}|$ and $\tilde{u} = \hat{\boldsymbol{\alpha}} \cdot \hat{\mathbf{u}}$, where $\hat{\boldsymbol{\alpha}} = \boldsymbol{\alpha}/\tilde{\alpha}$. Also equations (8.13), (8.14) can be rewritten as

$$i\tilde{\alpha}(U - c)\hat{w} + \frac{\mathrm{d}\tilde{p}}{\mathrm{d}z} = 0, \qquad (8.18)$$

$$i\tilde{\alpha}\tilde{u} + \frac{\mathrm{d}\hat{w}}{\mathrm{d}z} = 0, \qquad (8.19)$$

respectively. Now it can be seen that the transformed eigenvalue problem (8.17)–(8.19), (8.15) has the same form as the original eigenvalue problem (8.11)–(8.15) in the special case of two-dimensional wave modes when $\beta = \hat{v} = 0$; it follows that the eigenvalue relation for a given basic flow has the form

$$\mathcal{F}(c, \tilde{\alpha}) = 0 \qquad (8.20)$$

for some function $\mathcal{F}$, the eigenvalues c depending on the wavenumbers only through the sum of their squares. Thus the relative growth rate αc_i is greatest for given total wavenumber $\tilde{\alpha}$ when $\beta = 0$, because $\alpha \leq \tilde{\alpha}$ with equality only when $\beta = 0$.

The essential physical result is that a three-dimensional mode is a wave propagating obliquely to the plane of the basic flow, and that only the component of the basic flow in the direction of the wave affects the growth of the wave, so that the growth rate is proportional to this component. It implies that to each unstable three-dimensional mode there corresponds a more unstable two-dimensional one. You can see this more clearly by defining θ as the angle between the wavenumber vector $\boldsymbol{\alpha} = (\alpha, \beta, 0)$ and the direction of the basic velocity, namely, the direction of the x-axis. Then $\theta = \arctan(\beta/\alpha)$ and the component of the basic velocity in the direction of the wavenumber vector is

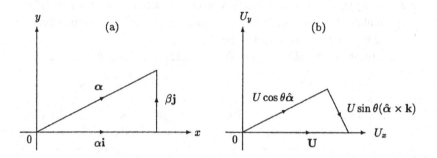

Figure 8.2   A sketch of the wavenumber vector and the components of the basic velocity in a plane $z = $ constant. (a) $\boldsymbol{\alpha} = \alpha\mathbf{i} + \beta\mathbf{j} = \tilde{\alpha}(\cos\theta\mathbf{i} + \sin\theta\mathbf{j})$. (b) $\mathbf{U} = U(\cos\theta\boldsymbol{\alpha} + \sin\theta\boldsymbol{\alpha} \times \mathbf{k})/\tilde{\alpha}$.

$U\cos\theta$, as shown in Figure 8.2. The equations show that the component of the basic velocity perpendicular to the wavenumber vector does not affect the growth of the wave.

Having just seen that the eigenvalue problem for a three-dimensional mode may be transformed into one for a two-dimensional mode, we need solve only the eigenvalue problem for two-dimensional modes. Rather than put $\beta = \hat{v} = 0$ in the eigenvalue problem above and eliminate $\hat{u}$, $\hat{p}$ in favour of $\hat{w}$, it is more convenient to start again, substituting a streamfunction ψ' of the perturbation in the linearized equations. So suppose that

$$u' = \frac{\partial\psi'}{\partial z}, \qquad v' = 0, \qquad w' = -\frac{\partial\psi'}{\partial x}, \qquad (8.21)$$

and take normal modes with

$$\psi'(x, z, t) = \phi(z)e^{i\alpha(x-ct)}. \qquad (8.22)$$

Therefore

$$\hat{u} = \frac{d\phi}{dz}, \qquad \hat{w} = -i\alpha\phi, \qquad (8.23)$$

and equation (8.11) gives

$$\hat{p} = \frac{dU}{dz}\phi - (U - c)\frac{d\phi}{dz}. \qquad (8.24)$$

On substituting this result for $\hat{p}$ into equation (8.13) we obtain the *Rayleigh stability equation*,

$$(U - c)(\phi'' - \alpha^2\phi) - U''\phi = 0, \qquad (8.25)$$

which, together with boundary conditions (8.15) in the form

$$\phi(z) = 0 \quad \text{at } z = z_1, z_2, \tag{8.26}$$

specifies the eigenvalue problem in its simplest terms, where we now use a prime to denote differentiation by z (you are unlikely to confuse it with the prime that denotes a perturbation quantity above). This is sometimes called the *Rayleigh stability problem*, being first posed by Rayleigh in 1880.

Note that the eigenvalue problem (8.25), (8.26) involves α only through α^2. Thus if ϕ, c are an eigensolution, that is, an eigenfunction and corresponding eigenvalue, for a given value of α, then they are also an eigensolution for the negative value $-\alpha$. This property is associated with the space and time reversibilities of the problem. On this basis, we shall henceforth take $\alpha \geq 0$ without loss of generality. Then a criterion for instability of a mode is that $c_i > 0$ and $\alpha > 0$. Further, note that the complex conjugate eigensolution $\phi^*, c^* = c_r - ic_i$ is also an eigensolution for the same wavenumber α because the eigenvalue problem is real. Therefore to each damped stable mode there is a corresponding amplified unstable mode, and *vice versa*. This is due to the time reversibility of the problem, which involves the periodic motions of an inviscid fluid with steady boundary conditions. The result may seem paradoxical, in the sense that if there is an exponentially decaying mode then there is also an exponentially growing mode, and that therefore the flow is unstable. It implies that the only way a flow may be stable is that *all* modes are neutrally stable with real eigenvalues c; but this is not surprising for a non-dissipative system, it being well known that for a Hamiltonian system of one degree of freedom all stable equilibria are centres and all unstable equilibria are saddles in the phase plane.

It can be seen that Rayleigh's stability equation (8.25) has a singularity at the point or points z_c in the domain of flow where $U(z_c) = c$ if $U_m \leq c \leq U_M$, where U_m is the minimum of $U(z)$ over the interval $z_1 \leq z \leq z_2$ of the flow, and U_M is the maximum. A plane where $z = z_c$ is called a *critical layer* of the mode. Note that a critical layer may occur only if c is real, that is, only if the mode is neutrally stable; however, critical layers do occur for marginally stable modes. It will be shown that critical layers are important in solving initial-value problems for an inviscid fluid, and in relating Rayleigh's stability problem to its generalization to a viscous fluid, namely the Orr–Sommerfeld problem.

In fact the eigensolutions c, ϕ for given smooth U, z_1, z_2, α are of two kinds (the modes for piecewise-linear profiles U are a little different). There is a *continuous spectrum* of eigenvalues c for *all* c in the interval $[U_m, U_M]$, each corresponding eigenfunction having a discontinuous derivative at z_c.

Also there *may* be a *discrete spectrum* of complex conjugate pairs of eigenvalues; the number of pairs is greater than or equal to zero and less than or equal to the number of inflection points of the velocity profile U.

8.2 General Properties of Rayleigh's Stability Problem

The work of the previous section is mostly due to Rayleigh (1880). He also found some general properties of the solution, and several explicit solutions of the eigenvalue problems for piecewise-linear basic velocity profiles. This section is a review of some general results found by him and others. Some details of specific solutions are given in the next section.

Perhaps Rayleigh's most famous and useful general result is that the occurrence of an inflection point in the basic velocity profile is a *necessary* condition for instability. His proof runs as follows. First rewrite Rayleigh's stability equation (8.25) as

$$\phi'' - \alpha^2 \phi - \frac{U''}{U - c}\phi = 0, \tag{8.27}$$

and suppose that the flow is unstable to *this* mode, so that $c_i > 0$. Now multiply equation (8.27) by the complex conjugate ϕ^* of ϕ, integrate from z_1 to z_2, integrate by parts, and use the boundary conditions (8.26). It follows that

$$\int_{z_1}^{z_2} \left(|\phi'|^2 + \alpha^2 |\phi|^2 \right) dz + \int_{z_1}^{z_2} \frac{U''}{U - c} |\phi|^2 \, dz = 0. \tag{8.28}$$

The imaginary part of equation (8.28) is

$$c_i \int_{z_1}^{z_2} \frac{U''}{|U - c|^2} |\phi|^2 \, dz = 0. \tag{8.29}$$

Now $c_i > 0$. Therefore U'' must change sign at least once in the open interval (z_1, z_2). This condition, being only a necessary one, cannot be used to show that any given flow is unstable, but can be used to show easily that some flows are stable.

Taylor (1915, pp. 23–26) explained the condition for instability physically by considering the transfer of x-momentum by the perturbation. He showed that if a perturbation is unstable and $U'' > 0$ throughout the fluid, then the x-momentum of every layer of the fluid must increase. But the momentum flux at the wall is zero because an inviscid fluid slips there, and so there is a contradiction.

He inferred that if the flow is unstable, then U'' is not positive throughout the fluid and, similarly, U'' is not negative throughout the fluid, thereby reaching

Rayleigh's result. It is interesting that Taylor added prophetically that the presence of an infinitesimal amount of viscosity might permit instability when U'' is of one sign throughout the fluid by permitting momentum transfer due to friction at a wall. Lin (1955, §4.4) has explained the physical mechanism of instability in terms of vorticity dynamics, identifying the point of inflection as a plane where the basic vorticity gradient vanishes, and Baines & Mitsudera (1994) in terms of a linear resonant mechanism. We shall come back soon to another physical explanation of Rayleigh's condition in terms of the Reynolds stress. However, the mechanism is rather complicated, and not readily intelligible in simple physical terms.

A stronger form of Rayleigh's condition was obtained 70 years later by Fjørtoft (1950): a *necessary* condition for instability is that $U''(U - U_s) < 0$ somewhere in the field of flow, where z_s is a point at which $U''(z_s) = 0$ and $U_s = U(z_s)$. The implications of Rayleigh's and Fjørtoft's conditions are illustrated in Figure 8.3.

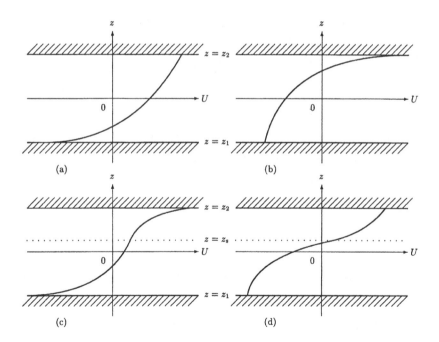

Figure 8.3 Some examples of flows governed by the Rayleigh–Fjørtoft necessary conditions for instability. (a) Stable because $U'' < 0$ everywhere. (b) Stable because $U'' > 0$ everywhere. (c) Stable because $U''(z_s) = 0$ but $U''(U - U_s) \geq 0$. (d) Possibly unstable because $U''(z_s) = 0$ and $U''(U - U_s) \leq 0$.

In practice, the Rayleigh–Fjørtoft conditions are often sufficient as well as necessary for instability. Tollmien (1935) showed plausibly that Rayleigh's condition is sufficient as well as necessary for instability of symmetric flows in a channel. His argument is based upon demonstrating the existence of a neutrally stable solution of the form

$$c = U_s, \qquad \alpha = \alpha_s > 0, \qquad \phi = \phi_s, \tag{8.30}$$

and then perturbing it to show that it is *marginally stable*, that is, there are both stable and unstable modes for values of α in each neighbourhood of α_s. Further, Lin (1955, p. 123) showed asymptotically that

$$c - U_s \sim -2\alpha_s \left\{ \int_{z_1}^{z_2} \phi_s^2 \, dz \right\} \Big/$$

$$\left\{ P \int_{z_1}^{z_2} \frac{U'' \phi_s^2}{(U - U_s)^2} \, dz + i\pi \, \text{sgn}[U']_{z=z_s} \left[\frac{U_s''' \phi_s^2}{U_s'^2} \right]_{z=z_s} \right\} \tag{8.31}$$

as $\alpha \to \alpha_s-$, where P denotes the Cauchy principal part of the integral. In fact there are unstable modes of this class for $0 < \alpha < \alpha_s$ but none for $\alpha > \alpha_s$.

Example 8.1: A counterexample to the sufficiency of the Rayleigh–Fjørtoft conditions for instability. Tollmien (1935) took the basic velocity profile

$$U(z) = \sin z \quad \text{for } z_1 \leq z \leq z_2.$$

Therefore $U'' = -U$ so that $U_s = 0$ if an integral multiple of π lies in the interval $[z_1, z_2]$ of flow, z_s is that multiple of π, and $U''(U - U_s) = -\sin^2 z \leq 0$. Therefore there is a marginally stable eigensolution with

$$c = U_s = 0, \qquad \alpha = \alpha_s = \left[1 - n^2 \pi^2 / (z_2 - z_1)^2 \right]^{1/2},$$

$$\phi = \phi_s = \sin[n\pi(z - z_1)/(z_2 - z_1)]$$

for each positive integer $n < (z_2 - z_1)/\pi$. It can thence be shown that if $z_2 - z_1 < \pi$ then the flow is stable, even though it has an inflection point at $z = 0$ within the domain of flow and Fjørtoft's necessary condition for instability is satisfied. □

Howard (1961) showed elegantly that if a mode is unstable, so that $c_i > 0$, then

$$\left[c_r - \tfrac{1}{2}(U_m + U_M) \right]^2 + c_i^2 \leq \left[\tfrac{1}{2}(U_M - U_m) \right]^2, \tag{8.32}$$

where U_m is the minimum of $U(z)$ over the interval $z_1 \leq z \leq z_2$ of the flow, and U_M is the maximum, that is, c lies in the semicircle in the upper half of the complex plane with centre $\frac{1}{2}(U_m + U_M)$ and radius $\frac{1}{2}(U_M - U_m)$. This inequality is called *Howard's semicircle theorem*. It is useful in giving easily an estimate of the growth rates and phase velocities of the normal modes, although it is formally a bound, not an estimate. Also it shows in the limit as $c_i \to 0+$ that $U_m \leq c_r \leq U_M$ for a marginally stable mode.

It has been noted that Rayleigh's stability equation has a singularity at the point or points z_c in the domain of flow where $U(z_c) = c$, that is, in the critical layers of the mode. The form of the streamlines near a critical layer was given by Kelvin (1880). On imposing a velocity equal to c on the whole system, that is, on making a Galilean transformation, the motion may be reduced to a steady flow relative to the transformed system. Then the streamlines are the same as the particle paths. The physical streamfunction for this steady flow is

$$\psi(x, z) = \Psi(z) + A\mathrm{Re}[\phi(z)\mathrm{e}^{\mathrm{i}\alpha x}],$$

where

$$\Psi(z) = \int_{z_c}^{z} [U(y) - c]\,\mathrm{d}y$$

is the streamfunction of the basic flow in the transformed system and A is a real constant proportional to the amplitude of the wave mode. Near the critical layer $z = z_c$ the equation of the streamlines is approximately

$$\tfrac{1}{2}U'(z_c)(z - z_c)^2 + A\phi(z_c)\cos\alpha x = \text{constant}, \tag{8.33}$$

where $\phi(z_c)$ is taken to be real by normalization. The streamlines now can be seen to have the famous *cat's-eye pattern* shown in Figure 8.4.

The Reynolds stress in the Reynolds–Orr energy equation (5.28) is the x-average

$$\tau = \frac{\alpha}{2\pi} \int_0^{2\pi/\alpha} (-w'u')\,\mathrm{d}x$$

$$= -\frac{\alpha}{2\pi} \int_0^{2\pi/\alpha} \mathrm{Re}(\hat{w}E)\mathrm{Re}(\hat{u}E)\,\mathrm{d}x,$$

where $E = \mathrm{e}^{\mathrm{i}\alpha(x-ct)} = F\exp(\alpha c_i t)$ and $F = \exp[\mathrm{i}\alpha(x - c_r t)]$. Therefore

$$\tau = -\frac{\alpha}{2\pi} \int_0^{2\pi/\alpha} \mathrm{Re}[(-\mathrm{i}\alpha\phi)E]\mathrm{Re}(\phi'E)\,\mathrm{d}x$$

$$= \tfrac{1}{4}\mathrm{i}\alpha(\phi\phi'^{*} - \phi^{*}\phi')\exp(2\alpha c_i t),$$

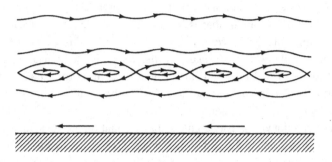

Figure 8.4   Kelvin's 'cat's-eye' pattern of the streamlines near the critical layer as viewed by an observer moving with the neutrally stable wave mode. (After Drazin & Reid, 1981, Fig. 4.3.)

because $\mathrm{Re}(\phi' E) = \frac{1}{2}(\phi' E + \phi'^* E^*) = \frac{1}{2}(\phi' F + \phi'^* F^{-1}) \exp(\alpha c_i t)$ and so forth. Now it follows that

$$\frac{d\tau}{dz} = \frac{1}{4}i\alpha\left(\phi\phi''^* - \phi^*\phi''\right)\exp(2\alpha c_i t)$$

$$= \frac{1}{2}\alpha c_i U''|U - c|^{-2}|\phi|^2 \exp(2\alpha c_i t), \qquad (8.34)$$

on using Rayleigh's stability equation (8.25). By consideration of a singularity at a critical layer where $U(z) = c$, it can be shown (see Exercise 8.4) that the 'jump' of the stress there is

$$\Delta\tau = \frac{1}{2}\pi\alpha\left[U''|\phi|^2/U'\right]_{z_c} \qquad (8.35)$$

in the limit as $c_i \to 0+$; this specifies the discontinuity of the Reynolds stress for a marginally stable mode. Here we denote the jump of the function τ at its discontinuity z_c by $\Delta\tau = \tau(z_c + 0) - \tau(z_c - 0)$.

These formulae have some important physical consequences. The Reynolds–Orr energy equation (5.28) shows that a perturbation extracts energy from (or gives energy to) the basic flow by means of the Reynolds stress, and so that the wave mode extracts energy where $\frac{1}{2}\tau U'$ is positive. Equation (8.34) shows that for a neutrally stable mode $d\tau/dz = 0$ wherever $U(z) \neq c$; therefore the Reynolds stress is constant between critical layers. Note also that τ vanishes at each wall because of the boundary condition that w vanishes there. This shows that the sum of the jumps of the Reynolds stress of a neutrally stable mode is zero. It follows that if there is at most one critical layer, as there must be if the basic velocity profile is monotonic, then the Reynolds stress must be zero throughout the flow. Therefore for a neutrally stable mode not only does $U'' = 0$ where $U = c$, but also $U = c$ where $U'' = 0$.

8.3 Stability Characteristics of Some Flows of an Inviscid Fluid

For plane Poiseuille flow and many channel flows, U'' does not vanish in the domain of flow, and so these flows *of an inviscid fluid* are stable to two-dimensional infinitesimal perturbations. But some channel flows, and all unbounded jets and free shear layers, are unstable. To solve the eigenvalue problem and hence find the stability characteristics for a specific basic flow, it is in general necessary to use numerical methods. However, a few results are known analytically, and these are useful pedagogically as examples.

Most of the known analytic results are for piecewise-linear basic velocity profiles. They are found by use of conditions satisfied by the perturbation at a discontinuity of U or U', which Rayleigh (1880) deduced and exploited. Suppose then that U or U' is discontinuous at $z = z_0$, say. Then it can be shown from equation (8.24) that continuity of pressure requires that

$$\Delta[(U - c)\phi' - U'\phi] = 0 \quad \text{at } z = z_0, \tag{8.36}$$

where we again use Δ to denote the jump of a quantity, so condition (8.36) means that $(U - c)\phi' - U'\phi$ is continuous at z_0. Also the continuity of the fluid at a perturbed interface, say $z = z_0 + \zeta(x, t)$, can be shown to give $D\zeta/Dt = w$ at $z = z_0 + \zeta$ and thence, on linearization,

$$\Delta\left[\frac{\phi}{U - c}\right] = 0 \quad \text{at } z = z_0. \tag{8.37}$$

Now, armed with conditions (8.36), (8.37), we are ready to solve any problem with a piecewise-linear velocity profile U, because in a layer where $U'' = 0$, Rayleigh's stability equation (8.25) has the general solution

$$\phi(z) = Ae^{\alpha z} + Be^{-\alpha z} \tag{8.38}$$

in simple explicit terms, for arbitrary constants A, B.

Example 8.2: Stability of plane Couette flow. See Figure 8.5(a). If

$$U(z) = z \quad \text{for } -1 \le z \le 1,$$

then Rayleigh's stability equation becomes

$$(z - c)(\phi'' - \alpha^2\phi) = 0,$$

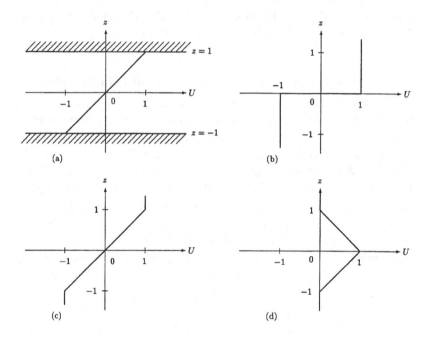

Figure 8.5   Sketches of some basic velocity profiles. (a) Plane Couette flow: $U(z) = z$. (b) Unbounded vortex sheet: $U(z) = \mathrm{sgn}(z)$. (c) An unbounded shear layer. (d) Triangular jet.

and therefore

$$\phi'' - \alpha^2 \phi = 0$$

wherever $z \neq c$. There is no solution of this equation which satisfies both of the boundary conditions (8.26), so the discrete spectrum is empty and plane Couette flow of an inviscid fluid is stable. However, the problem can be satisfied piecewise by continuous ϕ with a discontinuous derivative at the critical layer $z = c$ for all c such that $-1 < c < 1$ and for all α. The eigenfunction can readily be seen to be an arbitrary multiple of ϕ where

$$\phi(z) = \begin{cases} \dfrac{\sinh \alpha(1 + z)}{\sinh \alpha(1 + c)} & \text{for } -1 \leq z \leq c, \\[2ex] \dfrac{\sinh \alpha(1 - z)}{\sinh \alpha(1 - c)} & \text{for } c \leq z \leq 1. \end{cases}$$

This gives the continuous spectrum (after Case, 1960).   □

Example 8.3: Kelvin–Helmholtz instability. Helmholtz (1868) described the physical nature of the instability of a vortex sheet in a few words, and Kelvin (1871) analysed its linear instability fully, taking the basic velocity

$$U(z) = \begin{cases} 1 & \text{for } z > 0, \\ -1 & \text{for } z < 0, \end{cases}$$

as illustrated in Figure 8.5(b) and described in Chapter 3. Kelvin essentially solved Rayleigh's stability equation piecewise, using the boundary conditions at infinity, to deduce that

$$\phi(z) = \begin{cases} Ae^{-\alpha z} & \text{for } z > 0, \\ Be^{\alpha z} & \text{for } z < 0 \end{cases}$$

for some constants A, B. (Recall that we agreed to take $\alpha > 0$.) Now condition (8.36) applied at $z = 0$ gives

$$B/(-1 - c) = A/(1 - c),$$

and condition (8.37) gives

$$-\alpha(-1 - c)B = -\alpha(1 - c)A.$$

Eliminating A, B from these two linear homogeneous equations, we deduce that

$$c^2 = -1,$$

in agreement with equation (3.28). Therefore $c = \pm i$, and the relative growth rate $\alpha c_i = \pm \alpha$, giving one exponentially damped mode and one amplified mode for each value of the wavenumber. Therefore the flow is unstable to waves of all lengths.

Note that the growth rate tends to infinity as the wavelength tends to zero, so that the fine structure of an initial perturbation grows rapidly. To consider the effects of this rapid growth of short waves, suppose as an illustrative example (see Saffman, 1992, Chap. 8) that the initial conditions are such that the interfacial profile ζ has period $2\pi/\alpha$ and only the amplified modes are excited, so that

$$\zeta(x, t) = \sum_{n=1}^{\infty} A_n(t) \sin n\alpha x,$$

where $A_n(t) = A_n(0)e^{n\alpha t}$. Suppose further that $A_n(0) = \exp(-n^{1/2} - n\alpha t_0)$, say, for some $t_0 > 0$. It follows that ζ is smooth for $0 \leq t < t_0$ but not for

$t > t_0$. Thus a singularity of the curvature of the vortex sheet may develop in a finite time – at least according to this linear theory. □

Example 8.4: An unbounded continuous piecewise-linear shear layer. Take

$$U(z) = \begin{cases} 1 & \text{for } z \ge 1, \\ z & \text{for } -1 \le z \le 1, \\ -1 & \text{for } z \le -1, \end{cases} \tag{8.39}$$

as illustrated in Figure 8.5(c). We solve Rayleigh's stability equation piecewise, using the boundary conditions at infinity, to deduce that

$$\phi(z) = \begin{cases} A e^{-\alpha(z-1)} & \text{for } z > 1, \\ B e^{-\alpha(z-1)} + C e^{\alpha(z+1)} & \text{for } -1 < z < 1, \\ D e^{\alpha(z+1)} & \text{for } z < -1, \end{cases} \tag{8.40}$$

for some constants A, B, C, D. We both choose the factor e^α of the terms somewhat arbitrarily, and take $\alpha > 0$ as before, to simplify the algebra a little. Now the conditions (8.36), (8.37) applied at $z = \pm 1$ give

$$A = B + C e^{2\alpha}, \quad -(1-c)\alpha A = (1-c)\alpha\left(-B + C e^{2\alpha}\right) - \left(B + C e^{2\alpha}\right),$$

$$B e^{2\alpha} + C = D, \quad (-1-c)\alpha\left(-B e^{2\alpha} + C\right) - \left(B e^{2\alpha} + C\right) = \alpha(-1-c)D,$$

respectively. It is next a simple matter to eliminate A, B, C, D from these four linear homogeneous equations, by finding that

$$0 = \begin{vmatrix} 1 & 1 & e^{2\alpha} & 0 \\ -(1-c)\alpha & (1-c)\alpha(-1)-1 & [(1-c)\alpha-1]e^{2\alpha} & 0 \\ 0 & e^{2\alpha} & 1 & 1 \\ 0 & -[(-1-c)\alpha+1]e^{2\alpha} & (-1-c)\alpha-1 & \alpha(-1-c) \end{vmatrix}$$

and thence deducing the eigenvalue relation

$$c^2 = \left(4\alpha^2\right)^{-1}\left[(1-2\alpha)^2 - e^{-4\alpha}\right]. \tag{8.41}$$

This gives a pair of neutrally stable waves propagating in opposite directions if $c^2 > 0$ and a pair of stationary modes, one amplified and one damped, if $c^2 < 0$. Taking α_s as the (unique) positive zero of $1 - 2\alpha + e^{-2\alpha}$, we may calculate that $\alpha_s \approx 0.64$ and deduce that the mode is unstable if and only if $0 < \alpha < \alpha_s$. It follows that the shear layer is unstable. □

Example 8.5: A triangular jet. Take the basic velocity profile

$$U(z) = \begin{cases} 0 & \text{for } z \geq 1, \\ 1 - |z| & \text{for } -1 \leq z \leq 1, \\ 0 & \text{for } z \leq -1, \end{cases} \quad (8.42)$$

as illustrated in Figure 8.5(d). Again we will solve Rayleigh's stability equation piecewise, using the boundary conditions at infinity. It can be shown that if U is an even function and the boundary conditions are symmetric in $\pm z$, as in this problem, then each eigenfunction ϕ is either even or odd, so it is sufficient to consider the cases of even and odd eigenfunctions separately (see Exercise 8.13). The perturbation with even ϕ is called a *sinuous mode* or *antisymmetric mode*, because the jet oscillates sinusoidally and the streamlines are antisymmetric about the line $z = 0$, and that with odd ϕ is called a *varicose mode* or *symmetric mode*, because the jet looks varicose and the streamlines are symmetric about $z = 0$ (see Figure 8.6). Then, taking first the case of an even eigenfunction, suppose that

$$\phi(z) = \begin{cases} Ae^{-\alpha(|z|-1)} & \text{for } |z| > 1, \\ B\dfrac{\cosh \alpha z}{\cosh \alpha} + D\dfrac{\sinh \alpha |z|}{\sinh \alpha} & \text{for } |z| < 1, \end{cases}$$

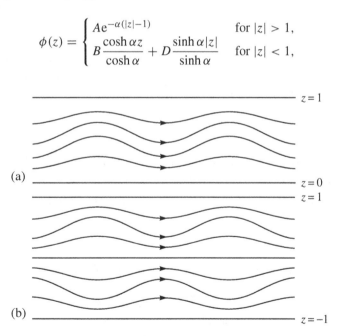

Figure 8.6   Symbolic sketches of modes of instability of a symmetric jet. They represent streamlines in a moving frame at a given instant. (Remember that modes travel at a constant velocity as they grow exponentially, are arbitrarily normalized, and are superposed on the basic flow.) (a) A sinuous mode. (b) A varicose mode.

for some constants A, B, D. Now the conditions (8.36), (8.37) applied at $z = \pm 1$ give

$$A = B + D, \qquad c\alpha A = -c\alpha(B\tanh\alpha + D\coth\alpha) + (B + D),$$

and applied at $z = 0$ give

$$(1 - c)\alpha D\operatorname{cosech}\alpha + B\operatorname{sech}\alpha = -(1 - c)\alpha D\operatorname{cosech}\alpha - B\operatorname{sech}\alpha.$$

Finally, elimination of A, B, D from these three linear homogeneous equations gives the eigenvalue relation

$$2\alpha^2 c^2 + \alpha\left(1 - 2\alpha - e^{-2\alpha}\right)c - \left[1 - \alpha - (1 + \alpha)e^{-2\alpha}\right] = 0. \qquad (8.43)$$

By solving this quadratic for c to test when there is a complex conjugate pair of roots, it can be shown that $\alpha_s = 1.833$, for which $U_s = 0.367$. Thus the jet is unstable to propagating growing sinuous waves which are not short.

Similarly, for the varicose modes it can be shown that

$$c = (2\alpha)^{-1}\left(1 - e^{-2\alpha}\right), \qquad (8.44)$$

by assuming that the eigenfunction is odd. It follows that the varicose mode is stable for all wavenumbers α. Note that the sinuous mode is more unstable than the varicose mode for this symmetric flow, and in fact this result is true much more generally. □

8.4 Nonlinear Perturbations of a Parallel Flow of an Inviscid Fluid

The weakly nonlinear theory of instability of a parallel flow of an inviscid fluid has yet to have a sound foundation built, because a given basic flow is either stable or unstable, and is not weakly unstable when a parameter, such as the difference between the value of the Reynolds number and its critical value, is small. But there is a lot more to be said.

Nonlinear Kelvin–Helmholtz instability is the problem of the rolling up of a vortex sheet between half spaces of irrotational flow, a classic problem of fluid dynamics. The singularity of the curvature of the interface at a finite time predicted by the linear theory of two-dimensional motion does in fact occur in the strongly nonlinear theory (see Saffman, 1992, Chap. 8). However, Beale *et al.* (1984) proved that for *two*-dimensional motion of an inviscid incompressible fluid, if the initial conditions are smooth then the velocity field is smooth for all time.

Here note that it is well known that in two-dimensional motion of an incompressible inviscid fluid the vorticity is convected with the fluid particles, so that the integral over the domain of flow of *any* given smooth function of the vorticity is a constant of the motion if no vorticity enters or leaves the domain. The kinetic energy is also a constant of the motion. It follows that in *two-dimensional flow*, instabilities in general develop smaller and smaller scales of motion.

Now the linear problem has been reduced by use of Squire's transformation essentially to one of two-dimensional vorticity dynamics. It is well known (Herivel, 1955) that such flows may be represented by a Hamiltonian system of infinite dimension, and that there is no attractor of a Hamiltonian system. (This system differs from the Hamiltonian system which describes the two-dimensional motion of a fluid particle along a streamline.) Also, from the point of view of the theory of dynamical systems, any steady flow may be regarded as either a generalized centre when stable or a generalized saddle point when unstable. From this point of view, it is natural that a steady flow has both growing and decaying perturbations when unstable but only neutrally stable perturbations when stable. This gives us some qualitative insight into the strongly nonlinear development of perturbations of a parallel flow of an inviscid fluid. If there is a unique mode of instability, then a general small initial perturbation will grow close to the one-dimensional unstable manifold of the basic flow; for periodic boundary conditions we usually find a unique unstable mode of an unstable parallel flow, but for an unbounded domain there is a *wave-band* of unstable modes.

However, it should be remembered that a model of an inviscid fluid is not structurally stable, so we expect weak damping with stable foci rather than centres when the fluid is slightly viscous, or even when there is weak dissipation due to truncation errors in modelling an inviscid fluid by use of computational fluid mechanics; then stable flows are attractors. For a real fluid the small scales of motion would eventually be diffused by viscosity, so that the small scales are averaged out locally in some way.

There is evidence from laboratory and numerical experiments that two-dimensional instabilities of some parallel flows grow such that the flow eventually becomes another two-dimensional steady flow which is periodic in the horizontal coordinate x. Also some vortices (see Exercise 8.17) have been found from a simple exact solution of the two-dimensional vorticity equation which is a strongly nonlinear generalization of the linear stability of an unbounded shear layer, but their significance is yet to be well understood.

It seems plausible, on the basis of laboratory and numerical experiments, that an initial perturbation of a basic parallel flow of a slightly viscous fluid grows in accordance with the theory of an inviscid fluid, so that after a while the flow is nearly periodic in space, with the fundamental wavelength of the most unstable mode, and that after a long time the viscosity 'smears' out the fine structure of the perturbation so that a steady two-dimensional flow of the given periodicity is reached approximately. This steady wavy flow appears to be essentially independent of what the small initial perturbations of the parallel flow are, stable to two-dimensional perturbations, and determined by maximizing some generalized form of the entropy, on using some arguments of statistical mechanics due to Onsager (1949). After an even longer time, three-dimensional instabilities break down this equilibrated wavy flow, and turbulence finally ensues.

Part 2: Viscous Fluid

8.5 Stability of Plane Parallel Flows of a Viscous Fluid

Again take

$$\mathbf{U} = U(z)\mathbf{i}, \qquad P = p_0 - Gx \quad \text{for } z_1 \le z \le z_2, \qquad (8.45)$$

to represent a basic flow along the channel with some constant adverse pressure gradient G. For a *viscous* fluid it can be verified (as in Example 2.4) that this flow satisfies the governing Navier–Stokes equations only if $d^3 U/dz^3 = 0$, that is, only for plane Couette–Poiseuille flow with a parabolic velocity profile.

Again, we may choose dimensionless variables, define a Reynolds number $R = VL/\nu$ in terms of some scales L of length and V of velocity of the basic flow, take small perturbations (8.5) of the basic flow, linearize the Navier–Stokes equations, take normal modes (8.10), and pose an eigenvalue problem to give c in terms of $U, z_1, z_2, (\alpha^2 + \beta^2)^{1/2}, \alpha R$. This again leads to Squire's transformation, with a modified Reynolds number $\tilde{R} = \alpha R/\tilde{\alpha} \le R$, there being equality only for a two-dimensional wave. This gives the important result, called *Squire's theorem*, that to obtain the minimum critical value of the Reynolds number it is sufficient to consider only two-dimensional perturbations.

On this basis, we shall consider only two-dimensional perturbations and exploit the convenience of both a streamfunction for two-dimensional flow and the vorticity equation in terms of the streamfunction; but it should be remembered that three-dimensional perturbations must be considered in the solution of some problems, because they are physically important in some initial-value problems and some nonlinear interactions. Then, start with the vorticity equation in the form

$$\frac{\partial(\Delta\psi)}{\partial t} + \frac{\partial(\Delta\psi, \psi)}{\partial(x, z)} = \frac{1}{R}\Delta^2\psi, \tag{8.46}$$

and the boundary conditions of no penetration and no slip on rigid walls at $z = z_1, z_2$ give

$$\psi(x, z, t) = \text{constant}, \quad \frac{\partial\psi}{\partial z}(x, z, t) = 0 \quad \text{at } z = z_1, z_2. \tag{8.47}$$

Now, on taking the streamfunction $\psi = \Psi + \psi'$, where $\Psi = \int U(z)\,dz$ is the basic streamfunction and ψ' is the perturbation streamfunction, linearization of vorticity equation (8.46) leads to

$$\frac{\partial(\Delta\psi')}{\partial t} + U\frac{\partial(\Delta\psi')}{\partial x} - \frac{d^2U}{dz^2}\frac{\partial\psi'}{\partial x} = \frac{1}{R}\Delta^2\psi'. \tag{8.48}$$

Next take normal modes (8.22), and it follows that

$$\phi^{iv} - 2\alpha^2\phi'' + \alpha^4\phi = i\alpha R[(U - c)(\phi'' - \alpha^2\phi) - U''\phi], \tag{8.49}$$

where again a prime is used to denote differentiation with respect to z; this is called the *Orr–Sommerfeld equation*. Further, boundary conditions (8.47) give

$$\phi(z) = \phi'(z) = 0 \quad \text{at } z = z_1, z_2. \tag{8.50}$$

The eigenvalue problem (8.49), (8.50) is called the *Orr–Sommerfeld problem*. Its solution for the eigenvalues can be expressed in the form

$$\mathcal{F}(c, \alpha^2, \alpha R) = 0. \tag{8.51}$$

This problem is attributed to Orr (1907b, Art. 25) and Sommerfeld (1908) independently, although they in fact only posed it for plane Couette flow.

Although the Orr–Sommerfeld equation becomes singular at each critical layer in the limit as $R \to \infty$, it is nonsingular for each given value of R, so $\mathcal{F}$ is an integral function of its arguments if the interval of flow is bounded and U is analytic over the domain; Schensted (1960) has proved that then there is a countable infinity of eigenvalues and the corresponding set of eigenfunctions is complete, so that an initial perturbation can be expressed as a linear combination of the discrete normal modes. If the interval of flow is unbounded, then there is a continuous spectrum of neutrally stable eigenmodes in addition to a discrete spectrum (see Herron, 1987).

For given U, α, R it is usually found that there is no unstable eigenmode, or just one or two. It is useful to plot the marginal curves of stability of the unstable modes in the (α, R)-plane, that is, to solve the eigenvalue relation to find the functions $c(\alpha, R)$ for each of the unstable modes and plot the graphs of $c_i(\alpha, R) = 0$. It follows that the minimum value of R along the curves is the critical value R_c of the Reynolds number, such that if $R < R_c$ then all modes are stable, but if $R > R_c$ then at least one mode is unstable.

Note that the Orr–Sommerfeld equation was derived on the assumption that the basic flow is both a plane parallel steady flow and an exact solution of the Navier–Stokes equations, so strictly the Orr–Sommerfeld problem is only applicable to the instability of plane Couette–Poiseuille flow. However, Tollmien (1929) simply took the Orr–Sommerfeld problem for $z_1 = 0$, $z_2 = \infty$ where U is the velocity of Blasius's boundary layer on a flat plate. This was ultimately supported by confirmation of his results by the careful laboratory experiments of Schubauer & Skramstad (1947), but may be made plausible as follows.

So let us digress for a while, reviewing Blasius's similarity solution of the boundary-layer equations to describe the flow of a uniform stream past a flat semi-infinite plate at zero angle of incidence. As is reported in many textbooks (e.g. Batchelor, 1967, §5.8), Blasius's solution is given by the streamfunction,

$$\Psi_* = (2\nu V x_*)^{1/2} f(\zeta),$$

where V is the velocity of the ambient uniform free stream parallel to the flat plate $z_* = 0$, f is the solution of Blasius's nonlinear ordinary-differential boundary-value problem, and the similarity variable $\zeta = (V/2\nu x_*)^{1/2} z_*$ acts locally as the distance normal to the plate. Thus the boundary-layer thickness $\delta = (2\nu x_*/V)^{1/2}$ increases parabolically with distance x_* downstream. This gives the basic velocity,

$$U_* = V f'(\zeta), \qquad V_* = 0, \qquad W_* = (\nu V/2x_*)^{1/2}[\zeta f'(\zeta) - f(\zeta)].$$

This solution is a valid approximation far from the leading edge $x_* = 0$ of the plate, where the *global* Reynolds number $V x_*/\nu$ is large. Far from the leading edge, the streamlines are nearly parallel, diverging little as x_* increases by a boundary-layer thickness δ. Near the leading edge the effects of viscosity are important and the streamlines diverge significantly as x_* increases by δ. It is clear then that this flow is nearly parallel far downstream, but not near the leading edge, and that the thickness of the layer increases without bound as $x_* \to \infty$. Where the boundary layer is nearly parallel we can use δ as a transverse length scale, V as a velocity scale, and define a *local* Reynolds number $R = V\delta/\nu = (2Vx_*/\nu)^{1/2}$, although R is proportional to the square root of the global Reynolds number and of the distance x_* downstream. The traditional method of treatment of the stability of the boundary layer, due to Tollmien (1929), is to take a given value of x_*, find the longitudinal velocity $U_*(x_*, z_*)$ at that station, express the velocity as $U_*(\zeta) = VU(z)$, where $z = z_*/\delta$, neglect the relatively small transverse velocity W_*, because it is smaller than U_* by a factor of $R^{-1/2}$, and then solve the Orr–Sommerfeld problem for the velocity profile U. This is a plausible procedure provided that (i) the resultant solution of the Orr–Sommerfeld problem gives a large critical value R_c of R, to ensure that the basic flow is indeed nearly parallel and the transverse flow is negligible at the onset of instability, and (ii) the most unstable waves are not too long, to ensure that the velocity profile does not develop much along one wavelength of each of those modes. (We shall see later, from the results of solving the Orr–Sommerfeld problem, that these provisos are in fact satisfied for Blasius's boundary layer on a plate.) However, the procedure was criticized by Taylor in 1938, although observations of experiments on the instability made since the time of Taylor's criticism have by and large supported the theoretical results found by Tollmien. Perhaps it should be stated that Tollmien's method leads to finding *where* the instability of the boundary layer begins, rather than *whether* it begins, because $R = R_c$ gives the station $x_* = R_c^2\nu/2V$ where instability begins, the boundary layer being unstable downstream where $R > R_c$. Also the boundary-layer approximation for large R, for which the pressure is independent of the transverse coordinate z, is fundamentally different from the linearization which leads to the Orr–Sommerfeld equation. It might seem that the linearized partial differential equations are parabolic with coefficients dependent on x_*, and so that the variable x_* cannot be separated to admit the use of normal modes of the form (8.22). This idea has been used in the derivation of the *parabolized stability equations* which represent weak nonparallelism and curvature of the basic flow as well as upstream influence (Bertolotti *et al.*, 1992); these equations are a parabolic system which can be easily integrated downstream, although the Navier–Stokes equations which they approximate are elliptic.

*On the basis of the above ideas, the development in space of a *spatial* mode of given dimensional frequency ω_* may be found by use of a WKBJ-like approximation (Bouthier, 1973; Gaster, 1974). Let $A_*(x_*)$ be the complex amplitude of the mode, perhaps triggered in the wind tunnel by a vibrating ribbon or a loudspeaker, with

$$\psi'_*(x_*, y_*, t_*) = \text{Re}[A_*(x_*)\phi(y_*, R(x_*), \omega_*)\exp(i\omega_* t_*)]$$

in dimensional form, and approximate

$$\frac{dA_*}{dx_*} = i\alpha_*(x_*, \omega_*)A_*,$$

where α_* is the dimensional complex wavenumber and ϕ the eigenfunction of the most unstable (or least stable) mode of the locally parallel flow at station x_*, found by solving the Orr–Sommerfeld problem at that station. Such a mode is called a *local mode* because it is determined by an approximation to the nearly parallel flow locally at a given station.

There is a lot more which has been said and written on the issue of non-parallelism; however, it is a deep issue, which has not been fully resolved, and we shall not pursue it further here. We shall simply take the Orr–Sommerfeld problem for a variety of nearly parallel flows U and solve it, but will bear in mind the limitations of this approximation.

8.6 Some General Properties of the Orr–Sommerfeld Problem

It is possible now, with modern computers and numerical methods, to calculate quite easily the eigensolutions for any given value of the wavenumber and hence to plot the marginal curve for any given basic flow. In the next section, some numerical results for some typical basic flows are presented. In the following section, the results of experiments and nonlinear theory are presented. However, there is more to be written first, about the physical mechanisms of instability, various useful analytic results, the subtle asymptotic structure of the solutions at large values of the Reynolds number, and the associated difficulties in the numerical solution of the problem.

The instability of parallel flows of a viscous fluid is notoriously subtle, but the physical mechanisms may be crudely described thus. If the basic velocity profile has a point of inflection and the profile is unstable for an inviscid fluid, then the action of viscosity is *chiefly* stabilizing. The shear instability, seen in Rayleigh's problem, is damped at small enough values of the Reynolds number R when viscosity dissipates the energy of the perturbation more rapidly than the Reynolds stress generates the energy. However, at large values of R, viscosity may *destabilize* a small band of waves which are stable when R is infinite. Indeed, often a basic flow is unstable for large and moderate values of R, although stable for an inviscid fluid; then each wave of given length is stable if R is large enough, because only a diminishing band of long waves is unstable as R increases to infinity (you may see this more clearly if you look ahead to Figure 8.8(a)). Thus viscosity has a dual role: a stabilizing role due to its dissipation of energy, and a more subtle destabilizing role. Taylor (1915), as narrated in §8.2, suggested that viscosity's prevention of slip at the walls might permit generation of x-momentum of a perturbation, and thereby cause instability. Prandtl (1921, 1935 §28) proposed that viscosity would lead to the transfer of energy from the basic flow to the perturbation near the critical layer by means of the Reynolds stress, because it would change the phase of the velocity components u', w' and hence the average of their product over a wavelength. So viscosity changes the results of §8.2 on the Reynolds stress of marginally stable, or weakly unstable, modes, smoothing out the jumps of the Reynolds stress near critical layers, and making the Reynolds stress non-zero near the critical layer when there is only one critical layer. The sign of the Reynolds stress that results is usually such that energy is transferred from the basic flow to the perturbation, in which case instability often follows, but sometimes from the perturbation to the basic flow. More recently, Lindzen & Rambaldi (1986) has explained the mechanism of instability physically in terms of what is called *over-reflection*, and Baines *et al.* (1996) in terms of a linear resonance (see Exercise 2.18).

We shall examine the energy balance and the structure of the critical layer in a little more detail next, to illuminate Prandtl's mechanism and other points.

8.6.1 Energy

To find the energy equation of the perturbation, first multiply the Orr–Sommerfeld equation (8.49) by the complex conjugate ϕ^* of ϕ, integrate from $z = z_1$ to z_2, integrate by parts, and use boundary conditions (8.50) to deduce that

$$-\mathrm{i}\alpha Rc\left(I_1^2 + \alpha^2 I_0^2\right) = -\left(I_2^2 + 2\alpha^2 I_1^2 + \alpha^4 I_0^2\right) - \mathrm{i}\alpha R \int_{z_1}^{z_2} \left[U\,|\phi'|^2 \right.$$
$$\left. + \left(U'' + \alpha^2\right)|\phi|^2 + U'\phi'\phi^*\right] \mathrm{d}z, \qquad (8.52)$$

where

$$I_n^2 = \int_{z_1}^{z_2} |\phi^{(n)}|^2 \, \mathrm{d}z \quad \text{for } n = 0, 1, 2. \qquad (8.53)$$

Taking the real part of equation (8.52), we find

$$\alpha Rc_\mathrm{i}\left(I_1^2 + \alpha^2 I_0^2\right) = -\left(I_2^2 + 2\alpha^2 I_1^2 + \alpha^4 I_0^2\right) - \frac{1}{2}\mathrm{i}\alpha R \int_{z_1}^{z_2} U'(\phi'\phi^* - \phi^{*\prime}\phi) \, \mathrm{d}z, \qquad (8.54)$$

which is essentially the energy equation (5.28) of the perturbation of the basic parallel flow. The terms on the left-hand side represent the rate of increase of the kinetic energy of the perturbation, the first group of terms on the right-hand side represents the rate of dissipation of the perturbation due to the viscosity, and the last term represents the energy transfer from the basic flow to the perturbation by means of the Reynolds stress (see Exercise 8.26).

Next abandon temporarily the Orr–Sommerfeld equation and consider the set S of smooth functions ϕ which satisfy the boundary conditions (8.50) and the energy equation (8.54). Thus S includes all the eigenfunctions of the Orr–Sommerfeld problem among an infinity of others. This permits the proof of *sufficient* conditions for stability by showing that $\alpha c_i < 0$ for *all* functions in the set S, for small enough values of the Reynolds number R. The physical basis of this method is that if R is small enough, viscosity dissipates energy more rapidly than it can be generated by the Reynolds stress for *any* geometrically possible flow which conserves mass and satisfies the boundary conditions, and that therefore then the flow is stable. This is an example of the energy method, which is essentially an application of Liapounov's direct method of stability with a Liapounov functional rather than a function.

So for a proof we require

$$0 > \alpha R c_i \left(I_1^2 + \alpha^2 I_0^2 \right) = -\left(I_2^2 + 2\alpha^2 I_1^2 + \alpha^4 I_0^2 \right)$$

$$- \frac{1}{2} i\alpha R \int_{z_1}^{z_2} U'(\phi'\phi^* - \phi^{*'}\phi) \, dz$$

for all $\phi \in S$. Now define $q = \max |U'|$ and note that

$$\left| \int_{z_1}^{z_2} U'(\phi'\phi^* - \phi^{*'}\phi) \, dz \right| \leq 2q \int_{z_1}^{z_2} |\phi'||\phi| \, dz$$

$$\leq 2q I_0 I_1, \tag{8.55}$$

by the Cauchy–Schwarz inequality. Therefore it remains to prove that

$$I_2^2 + 2\alpha^2 I_1^2 + \alpha^4 I_0^2 > \alpha R q I_0 I_1$$

for all $\phi \in S$.

Synge (1938) derived and used this result to deduce specific sufficient conditions for the stability of bounded flows (with $z_1 = -1$, $z_2 = 1$). He showed that one set of sufficient conditions is that

$$\left(\eta^2 q \alpha R \right)^2 < 4\left(2\alpha^2 \eta^2 + \xi\eta - \xi^2 + 2\eta \right)\left(\alpha^4 \eta^2 + \xi - 1 \right)$$

for any real pair ξ, η such that

$$2\alpha^2 \eta^2 + \xi\eta - \xi^2 + 2\eta > 0, \qquad \alpha^4 \eta^2 + \xi - 1 > 0.$$

(For unbounded flows it is necessary to take $\xi = 0$, and then the conditions are much weaker.) Taking $\xi = \eta = 1/\alpha$, this yields

$$(qR)^2 < 8(1 - \alpha^2 + \alpha^3 + \alpha^4),$$

and taking $\xi = 0, \eta = 3/\alpha^2$ yields

$$(qR)^2 < 256\alpha^4/27.$$

These conditions for stability are rather weak. Joseph (1968, 1969) sharpened the inequalities, but even then the conditions are not very useful because the energy method takes no account of the mechanism of instability due to the critical layer, and so cannot give a sharp bound on the value of R_c.

8.6.2 Instability in the inviscid limit

Asymptotics have played an important role in the historical development of the theory of the Orr–Sommerfeld problem; they pose a formidable mathematical challenge, and they are essential for the proper understanding of various physical mechanisms in the Orr–Sommerfeld and related problems, even though the Orr–Sommerfeld problem can be solved numerically today without great difficulty. However, in this book it is inappropriate to describe the asymptotics with any detail or rigour, so next we only sketch the asymptotic theory.

In order to relate the Orr–Sommerfeld problem to Rayleigh's problem, it is natural to take the limit as $R \rightarrow \infty$ in the Orr–Sommerfeld equation to find Rayleigh's equation. We note at once that the Orr–Sommerfeld equation is of fourth order and is nonsingular for all values of R, whereas Rayleigh's equation is of second order and has a singularity wherever $U(z) = c$ in the complex z-plane. Taking the limit as $R \rightarrow \infty$, we may find two independent asymptotic solutions of the Orr–Sommerfeld equation, ϕ_1, ϕ_2 say, called the *inviscid solutions*, which satisfy Rayleigh's equation (8.25), except possibly at a critical layer.

Now the Orr–Sommerfeld equation has not two but four independent solutions, such that the general solution of the Orr–Sommerfeld equation

$$\phi = A_1\phi_1 + A_2\phi_2 + A_3\phi_3 + A_4\phi_4 \tag{8.56}$$

for some constants A_1, A_2, A_3, A_4. Heisenberg worked on this problem before turning his attention to even more important problems of physics; in a paper of 1924 based on his doctoral thesis, he found the other two solutions asymptotically. He found, by what was afterwards to be called the WKBJ approximation, the *viscous solutions* such that

$$\phi_3(z), \phi_4(z) \sim [U(z) - c]^{-5/4} \exp\left[\mp(\alpha R)^{1/2} Q(z)\right] \quad \text{as } \alpha R \to \infty \quad (8.57)$$

for fixed $z \neq z_c$, where

$$Q(z) = e^{\pi i/4} \int_{z_c}^{z} [U(z') - c]^{1/2} \, dz'. \tag{8.58}$$

The viscous solutions represent the perturbation's boundary layers of thickness of order of magnitude $R^{-1/2}$ at the walls. Heisenberg used these approximate solutions to deduce heuristically that viscosity may indeed destabilize a given basic flow at large values of R as well as stablilize it at small values of R by dissipation of energy; more precisely, he showed that if the basic flow is bounded and $\alpha = \alpha_s > 0$, then there is an unstable mode with that wavenumber for sufficiently large values of R. However, at least one of the inviscid solutions and both the viscous solutions are singular at a critical layer, and so cannot validly approximate the exact solutions of the nonsingular Orr–Sommerfeld equation there.

Tollmien (1929) recognized this limitation of the Heisenberg solutions and examined a different approximation to the solutions of the Orr–Sommerfeld equation, where both z is close to z_c, and R is large. Note that $U(z) - c \sim U'(z_c)(z - z_c)$ as $z \to z_c$. The two terms ϕ^{iv}, $i\alpha R(U - c)\phi''$ in the Orr–Sommerfeld equation may dominate the others and balance with one another near the critical layer as $R \to \infty$. This intuitive argument suggests that the truncated equation

$$\phi^{iv} = i\alpha R(U - c)\phi''$$

gives a good approximation to the solution of the Orr–Sommerfeld equation near the critical layer for large αR. It may be more formally shown at length, on defining

$$\xi = (z - z_c)/\epsilon, \qquad \epsilon = \left[i\alpha R U'(z_c)\right]^{-1/3},$$

that the Orr–Sommerfeld equation becomes

$$\frac{d^4\phi}{d\xi^4} = \xi \frac{d^2\phi}{d\xi^2}$$

in the limit as $\alpha R \to \infty$ for *fixed* ξ, by substituting the variable ξ for z and the parameter ϵ for αR in the Orr–Sommerfeld equation and thereafter taking the limit. You may verify that this is indeed so. Therefore $d^2\phi/d\xi^2$ is an Airy function (see also Exercise 8.36). This gives four independent solutions, which can be expressed in simple explicit terms of Airy functions and their integrals. The solutions oscillate locally like an Airy function. Tollmien used these asymptotic solutions together with numerical methods in the first successful solution of the Orr–Sommerfeld problem, although his methods are rather crude by modern standards. It can be seen at once that the thickness of a critical layer, of the order of magnitude of ϵ, namely $R^{-1/3}$, is much greater than the thickness, of order $R^{-1/2}$, of a boundary layer at a wall. Tollmien also showed that the correct branch of a singular inviscid solution in the limit as $\alpha R \to \infty$ can be found from an unstable solution of the Rayleigh stability equation in the limit as $c_i \to 0+$.

Thus it emerges from these asymptotic theories that for large values of the Reynolds number a wave perturbation behaves very differently in four kinds of region, any of which may be present or absent according to the given basic flow:

(1) If the basic flow is uniform at infinity, then the perturbation is irrotational there, such that if $U(z) \to U_\infty$ as $z \to \infty$, then

$$\phi(z) \sim A_1 e^{-\alpha z} \quad \text{as } z \to \infty.$$

(2) The perturbation is rotational as if of an inviscid fluid, except at infinity, in a critical layer, or in the boundary layer near a wall, such that

$$\phi(z) = A_1 \phi_1(z) + A_2 \phi_2(z).$$

(3) In a critical layer of thickness $O[(\alpha R)^{-1/3}]$ near z_c where $U(z_c) = c$, the perturbation is substantially affected by the viscosity, however large αR is.

(4) Also, in a boundary layer of thickness $O[(\alpha R)^{-1/2}]$ near z_1 or z_2, the perturbation is substantially affected by the viscosity, however large R is.

The occurrence of these regions, examples of what are called *decks* in *triple-deck theory*, depends upon the basic flow and the values of the Reynolds number and the wavenumber. For the example of Blasius's boundary layer, on the *upper branch* of the marginal curve the perturbation has five decks (an irrotational 'inviscid' flow at infinity, a critical layer, two regions of rotational 'inviscid' flow on each side, and a boundary layer near the wall), but on the *lower branch* it has three decks because there c is small enough that the critical layer reaches to the wall $z = 0$ and absorbs the thinner boundary-layer region; see Figure 8.7 for a sketch of the decks and Figure 8.8(a) for the branches of the marginal curve. In the present context, triple-deck theory only provides a different way of deriving the classic asymptotic results for the Orr–Sommerfeld problem, but it also provides a way to find the asymptotic properties of a slowly developing

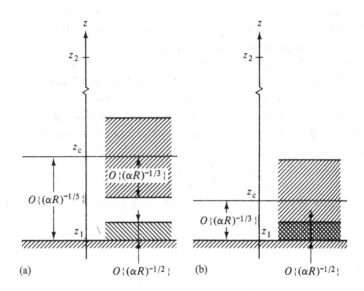

Figure 8.7   Sketches of the regions for a wave perturbation of the Blasius's boundary layer. The layer of irrotational flow at infinity is not depicted. (a) Quintuple deck for the upper branch of the marginal curve as $\alpha R \to \infty$. (b) Triple deck for the lower branch of the marginal curve. (After Drazin & Reid, 1981, Fig. 4.5.)

basic flow and weakly nonlinear perturbations for large values of R (Smith, 1979a,b).

It follows that the leading effects of viscosity for large values of R are due to a critical layer, if there is one, rather than a boundary layer at a wall. Within a critical layer a solution of the Orr–Sommerfeld equation is very different from any solution of Rayleigh's equation, and it is near the critical layer that the phases of u', w' are modified so that the Reynolds stress may transfer energy from the basic flow to the perturbation. In some cases the value of c is such that the critical layer reaches the wall and 'absorbs' the thinner boundary layer, and in other cases the critical layer is in the interior of the flow with an approximately inviscid region separating it from each of the boundary layers.

There is a lot more than the heuristic ideas of this section to be understood about the subtleties of the asymptotic theory of the solution of the Orr–Sommerfeld equation and its eigenvalue problem for large values of the Reynolds number. The above results about the asymptotic solutions of the Orr–Sommerfeld equation for fixed c become more complicated when the boundary conditions, which also determine c, are treated together with the asymptotic solutions. For this the reader is recommended to read Chapters 4 and 5 of Drazin & Reid (1981).

8.7 Stability Characteristics of Some Flows of a Viscous Fluid

The properties of the eigenvalue problem, which have complications beyond the scope of this book, may be summarized as follows at the risk of over-simplification. For a given basic velocity profile, (1) if a mode is unstable with eigenvalue c_∞ for an inviscid fluid ($R = \infty$), then its eigenvalue c for a viscous fluid tends to c_∞ as $R \to \infty$, but (2) if it is stable for an inviscid fluid, then c_∞ may not be the limit of any eigenvalue for a slightly viscous fluid. If the given flow is stable for $R = \infty$ (that is, stable to all modes when the fluid is inviscid), then it may be unstable, with at least one mode such that $c_i = O(R^{-1/3})$, as $R \to \infty$. Thus viscosity may render unstable a flow which is stable for an inviscid fluid; also it may make more unstable a flow which is unstable for an inviscid fluid. However, any given flow, unless it is an unbounded shear layer, is stable when the Reynolds number is sufficiently small. Note that, at large values of the Reynolds number, the growth rate of the Tollmien–Schlichting instability associated with the mechanism of the critical layer is much smaller than the growth rate of Rayleigh instability associated with the vorticity dynamics of an inviscid fluid. Some of the details to substantiate this summary will be explained in this section.

The Orr–Sommerfeld problem is not easy, so few solutions have been found analytically. It would seem that for a viscous fluid the basic velocity must be continuous and have continuous stress (so that the basic velocity gradient also is continuous if the fluid is of uniform viscosity). This is true, but nonetheless it makes some sense to consider the Orr–Sommerfeld problem when U or U' is discontinuous, as an approximation to the solution for long waves (small α). This is taken up in Exercise 8.39; here we shall solve the simplest example of all.

Example 8.6: Constant basic velocity. Suppose that U is constant in a channel. Then without loss of generality we may take

$$U(z) = 0 \quad \text{for } -1 \le z \le 1.$$

Now the Orr–Sommerfeld equation (8.49) becomes

$$\phi^{iv} - 2\alpha^2 \phi'' + \alpha^4 \phi = -i\alpha R c(\phi'' - \alpha^2 \phi).$$

The general solution of this equation can be written as

$$\phi(z) = A_1 \frac{\cosh \alpha z}{\cosh \alpha} + B_1 \frac{\sinh \alpha z}{\sinh \alpha} + A_2 \frac{\cos pz}{\cos p} + B_2 \frac{\sin pz}{\sin p}$$

for arbitrary constants A_1, B_1, A_2, B_2, where $p = (i\alpha Rc - \alpha^2)^{1/2}$. Again the sinuous and varicose modes can be separated. So for the sinuous mode (even ϕ) we take

$$\phi(z) = A_1 \frac{\cosh \alpha z}{\cosh \alpha} + A_2 \frac{\cos pz}{\cos p}.$$

Now boundary conditions (8.50) give

$$\phi(z) = A_1 \left(\frac{\cosh \alpha z}{\cosh \alpha} - \frac{\cos pz}{\cos p} \right),$$

where

$$\alpha \tanh \alpha + p \tan p = 0.$$

It can be shown (Rayleigh, 1892) that this eigenvalue relation has a countable infinity of real roots p, so $\alpha c = -i(\alpha^2 + p^2)/R$ has negative imaginary part, and all the sinuous modes are stable. It can be similarly shown (see Exercise 8.31) that all the varicose modes (with odd eigenfunctions) are stable, so this basic state of rest is stable. This is scarcely surprising because there is no Reynolds stress in the Reynolds–Orr equation (8.54) to generate instability.   □

In the 1960s, numerical calculations for plane Couette flow strongly suggested that it was stable to waves of all lengths at all values of the Reynolds number. This was proved by Romanov (1973). The first successful solutions of the Orr–Sommerfeld problem giving instability were found numerically by Tollmien (1929) and Schlichting (1933), in fact for the instability of Blasius's boundary layer on a flat plate. They used a method combining some asymptotic and numerical techniques that is now only of historical interest. However, modes of instability due to the critical-layer mechanism (rather than Rayleigh's inviscid mechanism) are still called *Tollmien–Schlichting waves* in their honour. The Orr–Sommerfeld equation is very stiff at large values of the Reynolds number (and these are usually the values of physical interest) and so not easy to solve numerically, and its difficulties were overcome only by use of electronic computers. The first solution of the Orr–Sommerfeld problem by use of an electronic computer was in 1953, when Thomas, at the suggestion of von Neumann, solved the problem for plane Poiseuille flow by direct numerical integration, and found instability at a few pairs of values of α, R. Now it is a routine matter to solve the Orr–Sommerfeld problem by any one of many good numerical methods (see Drazin & Reid, 1981, §30), although the stiffness of the equation still demands some care in the solution.

We shall give few details of results for particular basic flows, merely summarizing the main points of the many numerical solutions known now. The marginal curves sketched in Figure 8.8 and the numerical results in Table 8.1 do this very effectively, albeit briefly. First look at Figure 8.8(a). It is a sketch of a marginal curve typical of that for a basic flow which is stable for an inviscid fluid, such as plane Poiseuille flow and Blasius's boundary layer (but not plane Couette flow, because it happens to be stable for all values of the Reynolds number R). For these flows there is a unique unstable mode (the sinuous mode for plane Poiseuille flow). You can see that that there is no instability

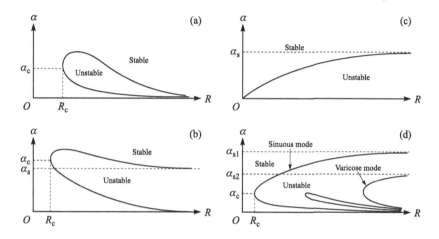

Figure 8.8   Sketches of some typical marginal curves of instability $c_i(\alpha, R) = 0$ in the (R, α)-plane for various classes of basic flows. (a) Basic flow which is stable when the fluid is inviscid. (b) Bounded flow which has one point of inflection. (c) Unbounded shear layer. (d) Unbounded jet or wake. (One mode is sinuous and the other varicose if and only if the jet is symmetric.)

Table 8.1. *Summary of numerical results for a few important prototypes of basic parallel flows*

Type of basic flow	Specific profile	Stable?	R_c	α_c	Marginal curve of Fig.
Uniform	$U = \text{constant}$	Yes	∞	–	–
Plane Couette	$U = z$: $-1 \leq z \leq 1$	Yes	∞	–	–
Plane Poiseuille	$U = 1 - z^2$: $-1 \leq z \leq 1$	No	5772	1.02	8.8(a)
Blasius's boundary layer	$U = f'(z)$: $z \geq 0$	No	520	0.30	8.8(a)
Shear layer	$U = \tanh z$: $-\infty < z < \infty$	No	0	0	8.8(c)
Jet or wake	$U = \text{sech}^2 z$: $-\infty < z < \infty$	No	4.02	0.17	8.8(d)

for any given value of the wavenumber α as $R \to \infty$. However, there is instability for at least one value of α for all R greater than R_c, because viscosity *destabilizes* the flow. Nonetheless, when R is small enough the dissipation of energy by viscosity dominates the destabilizing effect, and the flow is stable to all infinitesimal perturbations if $R < R_c$. Figure 8.8(b) is for a basic channel flow which has one marginally stable mode U_s, ϕ_s for an inviscid fluid, and so has one point of inflection. It can be seen that the results for an inviscid fluid arise as $R \to \infty$, and that sufficiently large viscosity stabilizes the flow.

Figure 8.8(c) is for a unbounded basic shear layer, say $U(z) = \tanh z$, which has one marginally stable mode U_s, ϕ_s for an inviscid fluid, and so has one point of inflection. You may be surprised to see that the critical Reynolds number is zero, so that there are unstable modes however large the viscosity is, in apparent contradiction of Serrin's theorem; nonetheless, this paradoxical result is not relevant physically because shear layers are far from parallel when their Reynolds number is small. Figure 8.8(d) is for a unbounded basic jet or wake, such as the Bickley jet with $U(z) = \operatorname{sech}^2 z$; such a basic flow has two points of inflection and so two marginally stable modes U_s, ϕ_s for an inviscid fluid. Again, the eigenvalues for small Reynolds number are not relevant quantitatively because then a jet or wake is not nearly parallel.

The asymptotic theory of the eigenvalue relation as R tends to infinity is mathematically challenging and beyond the scope of a first course in the subject, so we will merely quote a few results. For symmetric flows in a channel such as plane Poiseuille flow, $\alpha \sim$ constant $\times R^{-1/7}$, $c \sim$ constant $\times R^{-2/7}$ as $R \to \infty$ along the lower branch of the marginal curve, but $\alpha \sim$ constant $\times R^{-1/11}$, $c \sim$ constant $\times R^{-2/11}$ as $R \to \infty$ along the upper branch of the marginal curve. However, for semi-bounded flows of boundary-layer type, $\alpha \sim$ constant $\times R^{-1/4}$, $c \sim$ constant $\times R^{-1/4}$ as $R \to \infty$ along the lower branch. The scalings for the upper branch depend on the pressure gradient. In a favourable pressure gradient, $\alpha \sim$ constant $\times R^{-1/6}$, $c \sim$ constant $\times R^{-1/6}$, while for the zero-pressure-gradient Blasius boundary layer, $\alpha \sim$ constant $\times R^{-1/10}$, $c \sim$ constant $\times R^{-1/10}$ as $R \to \infty$.

The pattern of all the eigenvalues is indicated in Figure 8.9. It can be seen how there are *wall modes*, sometimes called *Airy modes* or *A-modes*, for which $c_r \to 1$, $c_i = O[(\alpha R)^{-1/3}]$ (with their critical layers near the walls), and *centre modes*, sometimes called *Pekeris modes* or *P-modes*, for which $c_r \to 0$, $c_i = O[(\alpha R)^{-1/2}]$ as $\alpha R \to \infty$. In addition, there are *Schensted modes* or *S-modes*, for which $c_r \to \frac{2}{3}$, $c_i \sim -i[\pi(n+2)]^2\alpha/4R$ as $\alpha R/n \to 0$; these occur for all values of the Reynolds number, even large ones, if the number n of the mode is sufficiently large, because then the mode's velocity varies so rapidly with z that the *local* Reynolds number of the mode is small (see Example 8.6).

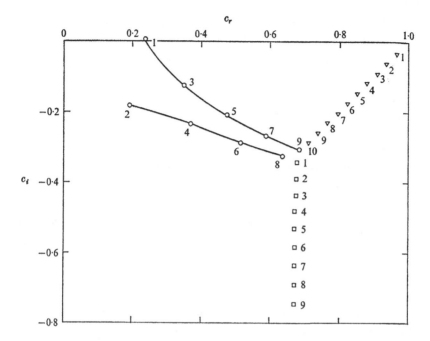

Figure 8.9   The first several eigenvalues $c = c_r + ic_i$ for sinuous modes of plane Poiseuille flow for $\alpha = 1$, $R = 10\,000$ plotted in the (c_r, c_i)-plane. (After Mack, 1976, Fig. 5.) Note the Airy $\circ$, Pekeris $\triangledown$ and Schensted $\square$ families of modes.

It is also interesting to see the eigenfunction and thence the nature of the perturbation for various velocity profiles at various pairs of values of the Reynolds number and wavenumber. The eigenfunction of the marginally stable mode for the Blasius boundary layer is shown in Figure 8.10. Note the structure of the mode and where the critical layer is.

8.8 *Numerical Methods of Solving the Orr–Sommerfeld Problem

We have remarked that it is a routine matter to solve the Orr–Sommerfeld problem by any one of many good numerical methods, although the stiffness of the equation still demands some care in the solution. The chief aims of the numerical solutions for a given basic flow are (1) to find the curve of marginal stability ($c_i = 0$) and curves of constant growth rate ($\alpha c_i = $ constant), (2) to find the eigenvalue spectrum for a given pair of positive values of α and R, and (3) to calculate the associated eigenfunctions and Reynolds stresses. There are similar aims for spatial modes of given real frequency $\omega = \alpha c$.

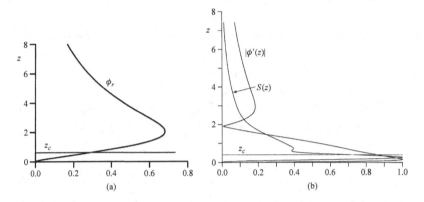

Figure 8.10   Sketches of an eigenfunction $\phi = \phi_r + i\phi_i$ for Blasius's boundary-layer profile on the upper branch, $R = 1000, \alpha = 0.23, c_r = 0.25, z_c = 0.40$. (a) $\phi_r(z)$ against z. (b) $|\phi'(z)|$ and $S(z) = \phi_r(z)\phi_i'(z) - \phi_r'(z)\phi_i(z)$ against z. Note that the Reynolds stress is proportional to S.

The chief methods of solution belong to two classes: (1) spectral expansion, i.e. expansion of ϕ as a linear combination of a complete set of orthogonal functions, such as Chebyshev polynomials, and (2) use of finite differences to integrate the Orr–Sommerfeld equation as an initial-value problem whereby integration is from one boundary towards the other. Various finite-difference schemes have been devised to deal with the stiffness of the equation for large values of R. Methods of spectral expansion are best suited to finding all the eigenvalues of the spectrum. Finite-difference methods are best suited to efficient calculation of the marginal curve (which involves only the most unstable eigenvalue) with use of a predictor and a corrector.

Drazin & Reid (1981, §20) summarize the technical details of many of these numerical methods and refer to many of the original papers. Schmid & Henningson (2001, Appendix) not only give an up-to-date summary but also list MATLAB programs to compute simply the stability characteristics of plane Couette and plane Poiseuille flows. These programs provide a useful platform on which students can base project work.

8.9 Experimental Results and Nonlinear Instability

Before the Second World War, the numerical solutions of the Orr–Sommerfeld problem for Blasius's boundary layer by Tollmien (1929, 1935) and Schlichting (1933) were not believed by all. The basic flow was not parallel, their asymptotics were heuristic, their calculations were arduous and complicated, and their results did not agree with experiments (or even with one another very precisely). However, during the war, Schubauer and Skramstad, by designing a wind tunnel with an unprecedentedly low level of turbulence upstream of the plate, by introducing controlled oscillations with a vibrating ribbon of desired frequencies and amplitudes, and by developing sensitive hot-wire anemometers to measure the growth and decay of the forced oscillations, were able to confirm the theoretical results about the 'nose' of the marginal curve quite convincingly. They published their results after the war (Schubauer & Skramstad, 1947).

Klebanoff *et al.* (1962) refined and developed these experiments. They found that first two-dimensional Tollmien–Schlichting waves grow in amplitude downstream, but where they reach a certain critical amplitude they become perturbed three-dimensionally and turbulent spots ensue due to apparently random bursting. The spatial evolution is illustrated schematically in Figure 8.11 and by a photograph in Figure 8.12. They also introduced controlled oscillations that varied spanwise to study the three-dimensionality, and interpreted the growth of three-dimensionality as a secondary instability of the primary Tollmien–Schlichting wave. An instantaneous velocity field exhibits large gradients in the mean velocity profile at spanwise positions

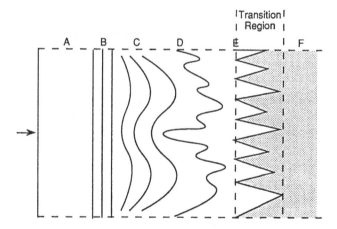

Figure 8.11   A symbolic sketch indicating roughly the regions of development of instability in Blasius's boundary layer on a plate at zero incidence in a low-turbulence stream: plan view. A, laminar flow; B, Tollmien–Schlichting (two-dimensional small-amplitude) waves; C, three-dimensional wave amplification; D, nonlinear peak-valley development with streamwise vortices; E, breakdown with formation and growth of turbulent spots; F, fully developed turbulence. (After Young, 1989, Fig. 5.13; reproduced by permission of Blackwell Science Ltd.)

Figure 8.12   Photographs of the development of instability of the boundary layer on a plate at zero incidence in a low-turbulence stream: flow from left to right. In the upper photograph there is laminar flow at a lower value of the Reynolds number, but Tollmien–Schlichting waves appear in the lower photograph at a five-times larger value of the Reynolds number. (After Van Dyke, 1982, Fig. 104; reproduced by permission of ONERA.)

(called *peaks*) where the secondary flow is directed away from the plate, and small gradients at intermediate spanwise positions (*valleys*); this velocity field corresponds to what are called Λ-*vortices*, shown in Figure 8.13.

It should be instructive to look now at all the relevant pictures of Van Dyke (1982, Figs. 29, 30, 104–106, 109–111) and of Nakayama (1988, Figs. 20, 21, 23, 24, 27). The film loops of Brown (FL1964) and Lippisch (FL1964) show the Tollmien–Schlichting waves developing in time as well as space. Also the *Video Library* of Homsy *et al.* (CD2000), under the subheadings 'Transition of Boundary Layers' and 'Visualization of Flow on a Flat Plate', and their *Boundary Layers*, under the subheadings 'Boundary Layer Flow', 'Flow Past a Sphere' and 'Instability, Transition, and Turbulence', have many short sequences of instabilities and transition to turbulence.

We have used Blasius's boundary layer on a plate here as a prototype of instability of parallel flows, but there have been many experiments on the

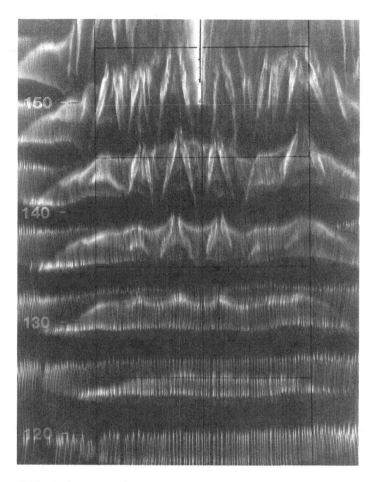

Figure 8.13 A photograph of Λ-vortices in instability of the boundary layer on a plate at zero incidence in a low-turbulence stream: plan view, with streaming from left to right. (After Saric; see Herbert, 1988, Fig. 3.)

instability of other parallel flows. Encouraging agreement between the linear theory and observations of carefully controlled small perturbations in wind and water tunnels with low levels of upstream turbulence has been found, but observations of subcritical instabilities and of instabilities triggered by higher levels of turbulence are less easy to interpret. Again, experimental agreement with the results of linear and weakly nonlinear theories of instability of plane Poiseuille flow (Nishioka *et al.*, 1980) leave no doubt about the validity of these theoretical models for describing careful experiments in low-turbulence wind tunnels and water channels.

The transition to turbulence is still imperfectly understood. The two-dimensional instability of the linear theory with Squire's transformation contrasts with the observed three-dimensionality of turbulence and also the local nature of turbulent spots. There are essentially two classes of theory to explain transition. For the first class, weakly nonlinear Tollmien–Schlichting waves develop secondary instabilities which are three-dimensional. There is a short sequence of bifurcations with successive regimes of flow downstream which lead rapidly to three-dimensionality, chaos and turbulence. For the second, there is some 'bypass' mechanism, whereby a subcritical perturbation of finite amplitude triggers turbulence directly.

The weakly nonlinear theories of Tollmien–Schlichting waves may further be classified in two, as those that treat the development and instability of a single wave, and those that treat the interaction of two or more waves. Following the pioneering work of Heisenberg (1924), Landau (1944) and Meksyn & Stuart (1951), the weakly nonlinear theory was initiated on a quantitative basis by Stuart (1960) and Watson (1960b), when they showed how to find the Landau equation governing the evolution of a weakly unstable perturbation of a parallel flow for small positive values of $R - R_c$. The Landau constant is in general complex, so that parallel flows are subject to Hopf bifurcations. Channel flows are in general subject to subcritical instability, and jets and unbounded shear layers to supercritical instability. However, for many flows the real part of the Landau constant has been found to change sign on the marginal curve of the linear theory. These have been confirmed by the careful experiments of Nishioka *et al.* (1975) on plane Poiseuille flow. Secondary instabilities of Tollmien–Schlichting waves lead to three-dimensionality in various weakly nonlinear theories, often with build-up of points of inflection in the mean velocity profile at some spanwise points of the flow, such that the ensuing instability (with negligible influence of viscosity) develops very rapidly. Orszag & Patera (1983) found numerically that a two-dimensional Tollmien–Schlichting wave superposed on a Blasius boundary layer on a plate was itself unstable. This secondary instability is three-dimensional, with regions of swirling flow shaped like an eccentric ellipse. It is believed (Bayly *et al.*, 1988) that this elliptical instablility is an important mechanism in the transition to turbulence of many flows.

Raetz (1959) considered the resonant interaction of three neutrally stable two-dimensional wave perturbations of a boundary layer. Kelly (1967) treated the subharmonic parametric instability of a shear layer. Craik (1971) treated the general case of the weakly nonlinear resonant interaction of three wave perturbations, two- or three-dimensional, of a parallel basic flow. He showed how this may lead to a subharmonic instability of boundary layers. His theory did not describe the experimental results of Klebanoff *et al.* (1962), who excited strong perturbations of chosen frequency. However, Saric & Thomas (1984) and Kachanov, Levchenko and co-workers (see Kachanov, 1994) did detect the predicted subharmonic resonance when they excited very weak perturbations of a boundary layer.

Since 1980 there has been an increasing number of direct numerical simulations of the nonlinear evolution of instabilities of parallel flows. Some of this work and its developments are described by Drazin & Reid (1981), much more by Schmid & Henningson (2001).

The relationship of modern experimental results to the weakly nonlinear theory is reviewed by Kachanov (1994); it might be said that there is good agreement with the theory and carefully controlled experiments with low levels of turbulence upstream, but the nature of transition in practice is less clear.

A crucial issue is that in practice transition in parallel and nearly parallel flows results from substantial perturbations upstream and irregularities on the surface of a channel and so forth, not from very carefully controlled forced sinusoidal perturbations of a flow with a very low turbulence level. One imaginative attempt to come to terms with this issue is a recent suggestion of a bypass mechanism by Butler & Farrell (1992) and Reddy & Henningson (1993) that the non-self-adjointness of the Orr–Sommerfeld problem results in very strong *transient* growth (perhaps by a factor of 10^5 or more when the Reynolds number is large) of a weakly unstable or even stable perturbation according to the *linear* theory. Exercise 8.44 gives an inkling of this mechanism by use of an analogy, but to understand the mechanism properly it is necessary to follow up the references. The mechanism intrinsically involves three-dimensional effects, and may give rise to nonlinearity after a period of strong amplification according to the linear theory.

Many aerodynamic engineers have set aside these complicated details of nonlinear instability in favour of the e^N *method*, proposed independently by van Ingen and by Smith & Gamberoni in 1956. This empirical method is the prediction that the onset of turbulence in a boundary layer occurs where a perturbation propagating downstream has grown, according to the linear theory, by a factor e^N, where $N = 9$ or thereabouts. It is clear that this method is conceptually flawed, because the magnitude of a perturbation at a given station is, at least according to the linear theory, proportional to its initial magnitude, and so the value of N must depend on the level of turbulence far upstream in the atmosphere or the wind tunnel. Also no nonlinear interaction or resonance is represented. However, as a rough-and-ready rule of thumb, the method is used widely with success.

This discussion has emphasized the instability and transition of Blasius's boundary layer on a flat plate at zero angle of incidence to a uniform stream. However, the instability and transition of other parallel and nearly parallel flows without a point of inflection, for example, channel flows and the boundary layer on a surface inclined to a uniform stream, are qualitatively similar.

Flows, like plane Couette flow, which are stable to all infinitesimal perturbations at all values of the Reynolds number, are fundamentally different. Even then, transition to turbulence due to small but finite-amplitude perturbations is somewhat similar to that of Blasius's boundary layer, at least similar in the sense of not yet being fully understood.

Flows with a point of inflection, or rather flows which are unstable at all large values of the Reynolds number, for example, a boundary layer with reversed flow, jets, wakes and free shear layers, are also fundamentally different. Their mechanism of instability is not Tollmien–Schlichting but the invisicid Rayleigh mechanism at large values of the Reynolds number, and sometimes a viscous mechanism at smallish values marking the onset of instability.

Schmid & Henningson (2001) describe at length many of the subtleties of transition of parallel flows, and cite a lot of the modern literature.

8.10 Stability of Axisymmetric Parallel Flows

In this section we shall review briefly the stability of some axisymmetric parallel flows, notably Poiseuille pipe flow and unbounded jets, of both inviscid and viscous fluids. The theory and results will be seen to be somewhat similar to those for plane parallel flows.

*It is appropriate to note first that the Orr–Sommerfeld problem is very special because the basic flow is not only steady and invariant under longitudinal translations but also two-dimensional. The invariance of a parallel flow under a longitudinal translation is associated both with the the separation of the longitudinal variable x in the stability problem and with the basic flow's independence of the Reynolds number. However, the linear stability problem for a given basic steady flow of a uniform viscous fluid is in general a non-separable partial differential system, and, if separable into an ordinary differential system, is typically of sixth order. (In the special case of plane parallel flow, the sixth-order system is factorizable into two independent systems, the Orr–Sommerfeld problem of fourth order and the Squire problem of second order – see Exercise 8.43.) These ideas are illustrated below by the problem for axisymmetric parallel flow.

Suppose then that there is a steady basic flow of the form

$$\mathbf{U}(\mathbf{x}) = U(r)\mathbf{i}, \qquad P(\mathbf{x}) = p_0 - Gx, \quad \text{for } r_1 \le r \le r_2, \tag{8.59}$$

between rigid cylinders at $r = r_1, r_2$ in terms of cylindrical polar coordinates (x, r, θ), where G is the imposed axial pressure gradient. The cylinders may be taken at rest or in uniform axial motion, and $r_1 = 0$ if there is no inner cylinder and $r_2 = \infty$ for unbounded flows such as jets. It can be verified that this solution satisfies the Navier–Stokes equations only if

$$U(r) = A + B \log r - Gr^2/4\rho\nu, \qquad (8.60)$$

for constants A, B, which are specified by the no-slip condition on the two cylinders. However, we may take other functions U to approximate the stability characterisics of nearly parallel flows of a viscous fluid, and any function U to get an exact solution of the Euler equations for an inviscid fluid.

Next formulate the stability problem after Sexl (1927a,b) and Batchelor & Gill (1962). First take small perturbations of this basic flow such that

$$\mathbf{u} = \mathbf{U} + \mathbf{u}', \qquad p = P + p', \qquad (8.61)$$

and linearize the governing Navier–Stokes equations. It is then possible to separate the variables, taking normal modes of the form

$$\mathbf{u}'(\mathbf{x}, t) = (\hat{u}_x(r), \hat{u}_r(r), \hat{u}_\theta(r)) \mathrm{e}^{\mathrm{i}(\alpha x + n\theta - \alpha ct)}, \qquad p'(\mathbf{x}, t) = \hat{p}(r) \mathrm{e}^{\mathrm{i}(\alpha x + n\theta - \alpha ct)},$$
$$(8.62)$$

for constant real axial wavenumber α and integral azimuthal wavenumber n. Note that the stability characteristics are the same if the sign of n is changed, because of the axisymmetry of the problem, so we will take $n \geq 0$ without loss of generality. It now follows, on taking dimensionless variables in the usual way, that

$$\hat{u}_x'' + \frac{1}{r}\hat{u}_x' - \left[\frac{n^2}{r^2} + \alpha^2 + \mathrm{i}\alpha R(U - c)\right]\hat{u}_x - RU'\hat{u}_r - \mathrm{i}\alpha R\hat{p} = 0, \quad (8.63)$$

$$\hat{u}_r'' + \frac{1}{r}\hat{u}_r' - \left[\frac{1+n^2}{r^2} + \alpha^2 + \mathrm{i}\alpha R(U - c)\right]\hat{u}_r - \frac{2\mathrm{i}n}{r^2}\hat{u}_\theta - R\hat{p}' = 0, \quad (8.64)$$

$$\hat{u}_\theta'' + \frac{1}{r}\hat{u}_\theta' - \left[\frac{1+n^2}{r^2} + \alpha^2 + \mathrm{i}\alpha R(U - c)\right]\hat{u}_\theta + \frac{2\mathrm{i}n}{r^2}\hat{u}_r - \frac{\mathrm{i}nR}{r}\hat{p} = 0, \quad (8.65)$$

$$\hat{u}_r' + \frac{1}{r}\hat{u}_r + \mathrm{i}\left(\alpha\hat{u}_x + \frac{n}{r}\hat{u}_\theta\right) = 0, \qquad (8.66)$$

where $R = VL/\nu$ is the Reynolds number and a prime denotes differentiation with respect to r. The boundary conditions of no penetration or slip of the perturbation on the cylinders give

$$\hat{u}_x(r) = \hat{u}_r(r) = \hat{u}_\theta(r) = 0 \quad \text{at } r = r_1, r_2 \qquad (8.67)$$

if $r_1 \neq 0$; but if $r_1 = 0$, the condition at $r = r_1$ is replaced by

$$\hat{u}_x(r) = \hat{u}_r(r) = \hat{u}_\theta(r) = \hat{p}(r) = 0 \quad \text{at } r = 0 \text{ for } n \neq 1, \qquad (8.68)$$

$$\hat{u}_x(r) = \hat{p}(r) = 0, \quad \hat{u}_r(r) + i\hat{u}_\theta(r) = 0 \qquad \text{at } r = 0 \text{ for } n = 1. \quad (8.69)$$

On defining the variables

$$k^2 = \alpha^2 + n^2/r^2, \quad \phi = -ir\hat{u}_r, \quad \Omega = (\alpha r\hat{u}_\theta - n\hat{u}_x)/k^2 r^2, \qquad (8.70)$$

and the linear operators

$$S = \frac{1}{k^2 r^3}\frac{d}{dr}\left(k^2 r^3 \frac{d}{dr}\right) - k^2, \qquad T = k^2 r \frac{d}{dr}\left(\frac{1}{k^2 r}\frac{d}{dr}\right) - k^2, \quad (8.71)$$

we may eliminate $\hat{u}_x$, $\hat{u}_r$, $\hat{u}_\theta$, $\hat{p}$ and deduce that

$$S\Omega + \left(2\alpha n/k^4 r^2\right)T\phi = i\alpha R\left[(U - c)\Omega - \left(nU'/\alpha k^2 r^3\right)\phi\right], \qquad (8.72)$$

$$T^2\phi - 2\alpha n T\Omega = i\alpha\left[(U - c)T\phi - k^2 r\left(U'/k^2 r\right)'\phi\right], \qquad (8.73)$$

where

$$\phi(r) = \phi'(r) = \Omega(r) = 0 \quad \text{at } r = r_1, r_2, \qquad (8.74)$$

or, if $r_1 = 0$ and $n \neq 0$,

$$\phi(r) = r^{2-n}\phi'(r) = \Omega(r) = 0 \quad \text{at } r = 0. \qquad (8.75)$$

It can be seen at once that if $n = 0$, then the equations (8.72), (8.73) decouple into a second-order system for Ω and a fourth-order system for ϕ, but if $n \neq 0$ they are a coupled sixth-order system.

For an inviscid fluid, we find similarly that

$$(U - c)T\phi - k^2 r\left(U'/k^2 r\right)'\phi = 0, \qquad (8.76)$$

or, if $n = 0$,

$$(U - c)L\phi - r(U'/r)'\phi = 0, \qquad (8.77)$$

where $L = d^2/dr^2 - r^{-1}d/dr - \alpha^2$. The boundary conditions in this case become

$$\phi(r) = 0 \quad \text{at } r = r_1, r_2. \tag{8.78}$$

The theory is closely analogous to that of the Rayleigh and Orr–Sommerfeld problems for plane parallel flows, and, indeed, Rayleigh (1880) initiated the theories for plane and axisymmetric parallel flows of an inviscid fluid in the same paper. The chief problems of physical interest are those of Poiseuille pipe flow and of an unbounded jet.

Numerical calculations (Salwen *et al.*, 1980) suggested strongly that Poiseuille pipe flow, with $U(r) = 1 - r^2$ for $0 \leq r \leq 1$, is stable to all *infinitesimal* perturbatons at all values of the Reynolds number. The spectrum of axisymmetric modes is somewhat similar to the spectrum for plane Poiseuille flow (see Figure 8.9), although no mode is unstable. The non-axisymmetric modes differ a little more (see Figures 8.14 and 8.15). In fact, transition to turbulence is observed experimentally for values of the Reynolds number down to about 2000 as the amplitude of perturbations in the inlet of the pipe is diminished (Reynolds, 1883).

The Landau–Squire exact similarity solution of the Navier–Stokes equations for a round jet is *locally* of the form $U(r) = 1/(1 + r^2)^2$ for $0 \leq r$ (see, e.g., Batchelor, 1967, p. 206). Treating the jet as an unbounded parallel flow, Burridge (1970) found numerically that it is unstable to modes for $n = 1$ whenever $R > R_c$, where $R_c \approx 37.5$ and $\alpha_c \approx 0.43$. From this and other

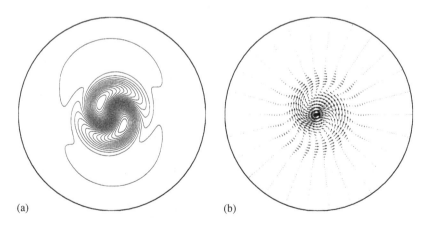

(a) (b)

Figure 8.14   A centre mode (the least stable mode) of Poiseuille flow for $R = 3000$, $n = 1, k = 1$. (a) Constant-z cross-sections of the streamfunction. (b) The vector field of the velocity perturbation. (After Meseguer & Trefethen, 2000, Fig. 4.)

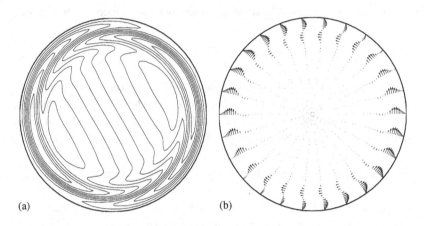

(a)   (b)

Figure 8.15   A wall mode of Poiseuille flow for $R = 3000$, $n = 1$, $k = 1$. (a) Constant-z cross-sections of the streamfunction. (b) The vector field of the velocity perturbation. (After Meseguer & Trefethen, 2000, Fig. 8.)

calculations, it seems that modes with $n = 1$ are usually the most unstable perturbations of axisymmetric parallel flows.

There are many calculations of the stability characteristics of axisymmetric flows of an inviscid fluid, ranging from Rayleigh (1880) to Batchelor & Gill (1962) to the present day. In particular, it is easy to use discontinuous velocity profiles with 'jump' conditions at their discontinuities analogous to those for plane parallel flows.

Exercises

8.1   *A derivation of the Rayleigh stability equation.* Taking the basic velocity $\mathbf{U}(z) = U(z)\mathbf{i}$ of an incompressible inviscid fluid, and using dimensionless variables, linearize the vorticity equation for two-dimensional flow,

$$\frac{\partial \eta}{\partial t} + u\frac{\partial \eta}{\partial x} + w\frac{\partial \eta}{\partial z} = 0,$$

where $\eta = \partial u/\partial z - \partial w/\partial x$, to show that

$$\frac{\partial \eta'}{\partial t} + U\frac{\partial \eta'}{\partial x} - \frac{d^2 U}{dz^2}\frac{\partial \psi'}{\partial x} = 0,$$

where $\eta' = \Delta\psi'$. Now, taking $\psi'(x, z, t) = \phi(z)e^{i\alpha(x-ct)}$, deduce the Rayleigh stability equation.

8.2 *The elevation of a material surface.* Show plausibly that if ζ is the elevation of a fluid particle above its basic position, say (x_0, y_0, z_0), when the basic flow $U(z)\mathbf{i}$ of an inviscid incompressible fluid is perturbed, then $w = D\zeta/Dt$ at $x = x_0, y = y_0, z = z_0 + \zeta$.

Deduce from the Rayleigh stability problem that the modes are of the form $\zeta(x, z, t) = F(z)e^{i\alpha(x-ct)}$, where $F = -\phi/(U - c)$ and

$$\left[(U - c)^2 F'\right]' - \alpha^2 (U - c)^2 F = 0. \qquad \text{(E8.1)}$$

8.3 *Some properties of the eigensolutions of the Rayleigh stability problem.* Define the inner product

$$\langle \phi, \psi \rangle = \int_{z_1}^{z_2} \phi(z)\psi(z)\,dz$$

over $C^2[z_1, z_2]$ and the linear operator L: $C^2[z_1, z_2] \to C[z_1, z_2]$ by

$$L\phi = (U - c)\left(\phi'' - \alpha^2\phi\right) - U''\phi.$$

Deduce that equation (E8.1), when divided by $U - c$, is the adjoint equation of the Rayleigh stability equation in the form (8.25), where F also satisfies the boundary conditions (8.26).

Multiplying equation (E8.1) by the complex conjugate F^* of F, integrating from z_1 to z_2, and using the boundary conditions, show that

$$\int_{z_1}^{z_2} (U - c)^2 \left(|F'|^2 + \alpha^2|F|^2\right) dz = 0.$$

Taking the imaginary part of this equation, deduce that if the flow is unstable to the mode, then c_r lies inside the range of U (that is, $U_m < c_r < U_M$, where $U_M = \max_{z_1 \leq z \leq z_2} U(z)$ and $U_m = \min_{z_1 \leq z \leq z_2} U(z)$). [Rayleigh (1880).]

8.4 *The jump of the shear stress at a critical layer of an inviscid fluid.* Show that

$$\frac{c_i U''|\phi|^2}{|U - c|^2} \sim \frac{c_i \left[U''|\phi|^2\right]_{z=z_c}}{U_c'^2 (z - z_c)^2 + c_i^2} \qquad \text{as } z \to z_c, c_i \to 0+ \qquad \text{(E8.2)}$$

if $U(z_c) = c_r, U_c' \neq 0, [U''|\phi|^2]_{z=z_c} \neq 0$. Deduce plausibly that

$$\lim_{c_i \to 0+} \lim_{\epsilon \to 0} \int_{z_c-\epsilon}^{z_c+\epsilon} \frac{c_i U''|\phi|^2}{|U - c|^2}\,dz = \pi \left[\frac{U''|\phi|^2}{U'}\right]_{z=z_c}.$$

Hence show that equation (8.35) follows from (8.34).

8.5 *Orr's 'lift-up' mechanism.* Taking plane Couette flow with basic velocity $\mathbf{U} = z\mathbf{i}$, show that the linearized vorticity equation for two-dimensional perturbations is

$$\frac{\partial \eta'}{\partial t} + z\frac{\partial \eta'}{\partial x} = 0,$$

in the usual notation, where $\eta' = \Delta\psi'$. Deduce that

$$\Delta\psi' = F(x - zt, z)$$

for an arbitrary differentiable function F.

Discuss the relationship of the Rayleigh stability problem for plane Couette flow to the above approach to the solution of the initial-value problem.

Discuss the above representation of the convection of vorticity of the perturbation by the basic flow. It is sometimes called the *lift-up mechanism* and sometimes the *Venetian-blind mechanism*. [Orr (1907a, Art. 13).]

8.6 *Stability or instability of plane Couette flow?* Reconsider Exercise 8.5 as an initial-value problem to determine η' over the strip $\mathcal{V} = \{(x, z): -\infty < x < \infty, -1 \leq z \leq 1\}$ in the (x, z)-plane as follows. First assume that F and $\partial F/\partial z$ are bounded and continuous in $\mathcal{V}$.

(i) Defining the norm $\|\eta\| = \sup_{\mathbf{x} \in \mathcal{V}} |\eta(x, z)|$, show that the null solution $\eta' = 0$ is stable.

(ii) Show, however, that the null solution $\eta' = 0$ is unstable if the norm is defined as $\|\eta\| = \sup_{\mathbf{x} \in \mathcal{V}} \{(\partial \eta(x, z)/\partial z)^2 + [\eta(x, z)]^2\}^{1/2}$. [Yudovich (1989, p. 101).]

8.7 **An initial-value problem for two-dimensional perturbations of plane Couette flow of an inviscid fluid.* Consider a basic flow of an inviscid incompressible fluid with velocity $\mathbf{U} = z\mathbf{i}$ between rigid planes at $z = \pm 1$ moving with the flow. Verify that an exact solution of the *nonlinear* Euler equations is given by the streamfunction perturbation

$$\psi'(x, z, t) = \frac{a\cos(\alpha x + \gamma z - \alpha t z)}{\alpha^2 + (\gamma - \alpha t)^2}$$

for $-\infty < x < \infty, -1 \leq z \leq 1$, and all real a, α, γ.

Show that the kinetic energy density of the wave perturbation per unit mass per wavelength,

$$K(t) = \frac{\alpha}{2\pi} \int_0^{2\pi/\alpha} dx \int_{-1}^1 dz \, \tfrac{1}{2}\left(u'^2 + w'^2\right)$$

$$= \frac{(1/2)a^2}{\alpha^2 + (\gamma - \alpha t)^2}.$$

Deduce that if the wave fronts are initially tilted in the same direction as the basic velocity profile (that is, γ has the same sign as α), then K decreases monotonically to zero as t increases. Deduce that otherwise (γ has the opposite sign) K increases until $t = \gamma/\alpha$ and thereafter decays to zero; show that its maximum is large for a wave with a short length in the direction of the basic flow.

Show further that if

$$\psi'(x, z, 0) = b \sin\left[\tfrac{1}{2}n\pi(z + 1)\right]e^{i\alpha x},$$

where n is a positive integer, then the *linearized* problem has a solution of the form

$$\psi'(x, z, t) = -\tfrac{1}{2}ib\left(\alpha^2 + \tfrac{1}{4}n^2\pi^2\right)e^{i\alpha x}$$

$$\times \left[\frac{e^{in\pi(z+1)/2 - i\alpha t z}}{\alpha^2 + \left((1/2)n\pi - \alpha t\right)^2} - \frac{e^{-in\pi(z+1)/2 - i\alpha t z}}{\alpha^2 + \left((1/2)n\pi + \alpha t\right)^2} \right.$$

$$\left. + A(t)\frac{\cosh \alpha z}{\cosh \alpha} + B(t)\frac{\sinh \alpha z}{\sinh \alpha} \right],$$

where A, B may be chosen so that ψ' satisfies the boundary and initial conditions. Show that

$$\psi'(x, z, n\pi/2\alpha) = -\tfrac{1}{2}i\left(\alpha^2 + \tfrac{1}{4}n^2\pi^2\right)be^{i(\alpha x + n\pi/2)}$$

$$\times \left[\frac{1}{\alpha^2} - \frac{e^{-in\pi(z+1)}}{\alpha^2 + n^2\pi^2} - \frac{\cosh \alpha z}{\cosh \alpha}\left(\frac{1}{\alpha^2} - \frac{1}{\alpha^2 + n^2\pi^2}\right) \right],$$

and thence that

$$\psi'(x, z, n\pi/2\alpha) \sim -\tfrac{1}{16}in^2\pi^2be^{i(\alpha x + n\pi/2)}\left(1 - z^2\right) \quad \text{as } \alpha \to \infty$$

for fixed z, n. Deduce further that, *in terms of the linear theory for an inviscid fluid*, an initial perturbation may become arbitrarily large after a long time although it ultimately decays like t^{-2} as $t \to \infty$. Is the flow stable in the sense of Liapounov? What do you expect would happen in the presence of nonlinearity? [Kelvin (1887, §§32–34), Orr (1907a), Craik & Criminale (1986, p. 18).]

8.8   *An apparent two-dimensional algebraic instability of a plane parallel Couette flow of an inviscid fluid.* Show that two-dimensional perturbations of a basic flow of an incompressible inviscid fluid with velocity $\mathbf{U} = U(z)\mathbf{i}$ satisfy the linearized equations,

$$\frac{\partial u'}{\partial t} + \frac{dU}{dz}w' = 0, \qquad \frac{\partial v'}{\partial y} + \frac{\partial w'}{\partial z} = 0, \qquad \frac{\partial \xi'}{\partial t} = 0,$$

where $\mathbf{u}'$, p' are assumed to be independent of x and where $\xi' = \partial w'/\partial y - \partial v'/\partial z$. Deduce that $u'(y, z, t) = u'(y, z, 0) - tU'(z)w'(y, z, 0)$, $v'(y, z, t) = v'(y, z, 0), w'(y, z, t) = w'(y, z, 0)$. [Ellingsen & Palm (1975). See Exercise 8.25.]

8.9   *Instability of parallel flows of an inviscid fluid.* Verify that

$$\mathbf{u} = U(y)\mathbf{i} + W(x - tU(y))\mathbf{k}, \qquad p = \text{constant},$$

satisfies Euler's equations of motion and the equation of continuity of an incompressible inviscid fluid. Deduce that the vorticity

$$\boldsymbol{\omega} = \nabla \times \mathbf{u} = -tU'(y)W'(x - tU(y))\mathbf{i} - W'(x - tU(y))\mathbf{j} - U'(y)\mathbf{k}.$$

Now consider the stability of the basic flow with solution $\mathbf{U} = U(y)\mathbf{i}$ for $-\infty < y < \infty$ to perturbations with period 2π in x and z. Defining the norm of a perturbation $\mathbf{u}'$, p' with period 2π in x and z as

$$\|\mathbf{u}'\| = \left[\int_0^{2\pi} dx \int_{-\infty}^{\infty} dy \int_0^{2\pi} dz \left(\boldsymbol{\omega}'^2 + \mathbf{u}'^2 \right) \right]^{1/2},$$

show that there exist small initial perturbations of the basic solution such that $\|\mathbf{u}'\| \sim \text{constant} \times t$ as $t \to \infty$ provided that U' is square-integrable and U not constant, and thence that the basic flow is unstable, whether or not it has a point of inflection.

What does this tell us about our choice of definition of stability, what about nonlinear perturbations, and what about the flow? [Yudovich (1989, p. 101).]

8.10 *Instability of a broken-line plane jet of an inviscid fluid.* Consider the broken-line jet of an incompressible inviscid fluid in an unbounded plane parallel flow with the velocity profile,

$$U(z) = \begin{cases} 0 & \text{if } z > 1 \\ 1 & \text{if } -1 < z < 1 \\ 0 & \text{if } z < -1. \end{cases}$$

(It is sometimes called a 'top-hat' jet, because of the shape of the graph of U.) Show that the sinuous modes have eigenvalue relation,

$$(c + 1)^2 \tanh |\alpha| + c^2 = 0.$$

Deduce that the jet is unstable to waves of all lengths.

Find the eigenvalue relation of the varicose modes. [Rayleigh (1894, pp. 380–381).]

8.11 *Instability of plane Couette flow between two free surfaces.* Consider the basic parallel flow of an incompressible inviscid fluid with velocity $U(z) = z$ for $-1 \le z \le 1$, where the surfaces with basic positions $z = \pm 1$ are free.

Show that ϕ satisfies the Rayleigh stability equation for $-1 < z < 1$ and $(U - c)\phi' - U'\phi = 0$ at $z = \pm 1$. Deduce that

$$c^2 = \frac{(\alpha \tanh \alpha - 1)(\alpha - \tanh \alpha)}{\alpha^2 \tanh \alpha},$$

and thence that the flow is unstable if $\alpha < \alpha_c$, where $\alpha_c \approx 1.2$ is defined as the positive root of $\alpha \tanh \alpha = 1$.

8.12 *Instability of a free shear layer.* Show that if $U(z) = \tanh z$ for $-\infty < z < \infty$, then $U_s = 0$. Verify that the corresponding marginally stable solution of the Rayleigh stability problem is

$$c = 0, \qquad \alpha = 1, \qquad \phi = \text{sech } z.$$

Show that

$$\left[\frac{dc}{d\alpha} \right]_{\alpha = 1-} = -\frac{2i}{\pi}.$$

Infer that the normal mode is unstable for $0 < \alpha < 1$.

8.13 *Symmetry: sinuous and varicose modes.* Suppose that S: $C^2[-L, L] \to C[-L, L]$ is such that $S f_e$ is an even function for all even $f_e \in C^2[-L, L]$, and $S f_o$ an odd function for all odd $f_o \in C^2[-L, L]$. Deduce that if

$$(Sf)(z) = 0 \quad \forall z \in [-L, L], \quad \text{and} \quad f(\pm L) = 0,$$

then

$$(S f_e)(z) = 0 \quad \forall\, z \in [-L, L], \qquad f_e(\pm L) = 0$$

and

$$(S f_o)(z) = 0 \quad \forall\, z \in [-L, L], \qquad f_o(\pm L) = 0,$$

where

$$f_e(z) = \tfrac{1}{2}[f(z) + f(-z)], \quad f_o(z) = \tfrac{1}{2}[f(z) - f(-z)] \; \forall\, z \in [-L, L].$$

[Note that f_e or f_o may be identically zero.]

Show that if the operator T is defined by $Tf = d^2 f/dz^2 + \lambda f$ for all $f \in C^2[-\pi, \pi]$, then T shares the symmetry properties of S above. Show further that the problem

$$-\frac{d^2\phi}{dz^2} = \lambda\phi, \qquad \phi(\pm\pi) = 0,$$

has eigensolutions $\lambda = \lambda_n, \phi = \phi_n$ for $n = 0, 1, \ldots,$ where $\lambda_n = \tfrac{1}{4}(n + 1)^2$,

$$\phi_{2m}(z) = \cos\left(m + \tfrac{1}{2}\right)z, \qquad \phi_{2m+1}(z) = \sin(m + 1)z.$$

8.14 *Instability of the Bickley jet.* Show that if $U(z) = \operatorname{sech}^2 z$ for $-\infty < z < \infty$, then $U_s = \tfrac{2}{3}$. Verify that the corresponding marginally stable solutions of the Rayleigh stability problem are the sinuous mode

$$c = \tfrac{2}{3}, \qquad \alpha = 2, \qquad \phi = \operatorname{sech}^2 z,$$

and the varicose mode

$$c = \tfrac{2}{3}, \qquad \alpha = 1, \qquad \phi = \operatorname{sech} z \tanh z.$$

Infer that the jet is stable to all short waves with $\alpha > 2$ and unstable to some long waves with $\alpha < 2$.

8.15 *The stabilizing effect of boundaries on a shear layer and on a jet of an inviscid fluid.* Consider the basic parallel flow of an incompressible inviscid fluid with velocity $U(z)$ between rigid boundaries at $z = \pm L$. You are given that the flow is unstable to some temporally growing infinitesimal perturbations if $L = \infty$ and is stable in the limit as $L \to 0$. Show that, to calculate the value of L_c such that the flow is unstable if

$L > L_c$ and stable if $L < L_c$, it is sufficient to put $\alpha_s = 0$ and solve the boundary-value problem

$$\phi_s'' - \frac{U''}{U - U_s}\phi_s = 0,$$

$$\phi_s = 0 \quad \text{at } z = \pm L_c.$$

Verify that $\phi_s = U - U_s$ is one solution of the equation.

(i) First suppose that $U(z) = \tanh z$. To what range of wavenumbers is the shear layer unstable when $L = \infty$? Why is the flow stable in the limit as $L \to 0$? [Hints: Exercise 8.12, Exercise 8.37, Example 8.2.]

Next calculate L_c. [Note: L_c is a root of a simple transcendental equation which need *not* be solved explicitly; a sketch showing that there exists a unique positive root and an estimate of L_c to one significant figure will suffice.]

(ii) Calculate the corresponding value of L_c when $U(z) = \operatorname{sech}^2 z$ for $-L < z < L$. Indicate why you may assume in the calculation that ϕ_s is an even function. Show that this symmetric jet with symmetric boundary conditions remains unstable so long as the points of inflection remain within the domain of flow. [Hint: Exercise 8.14.]

[After H. E. Huppert (private communication); Howard (1964).]

8.16 *Marginally stable modes of an unbounded jet.* Consider the Rayleigh stability problem numerically for the jet with $U(z) = \operatorname{sech} z$ for $-\infty < z < \infty$.

(i) Show that

$$\phi(z) \sim \text{constant} \times e^{\alpha z} \quad \text{as } z \to -\infty.$$

Deduce plausibly that the boundary condition that $\phi(z) \to 0$ as $z \to -\infty$ is well approximated by taking $\phi' - \alpha\phi = 0$ at $z = -L$ instead, where L is large, the approximation being better the larger L is.

(ii) Show that $U'' = (1 - 2U^2)U$, $U_s = 2^{-1/2}$, and thence that

$$\phi_s'' - [\alpha_s^2 - U(2U + 2^{1/2})]\phi_s = 0,$$

$$\phi_s(z) \to 0 \quad \text{as } z \to \pm\infty.$$

Show that $\phi_s'(0) = 0$ for the sinuous mode and $\phi_s(0) = 0$ for the varicose mode.

(iii) You may now compute the marginally stable sinuous mode by a 'shooting method' as follows. Your computing, from start to finish, should take less than an hour *if* you are already familiar with a suitable software system having a package which integrates ordinary differential equations.

To determine α_s for the sinuous mode, guess the value of α_s, take the system

$$\phi'_s = v, \qquad v' = [\alpha_s^2 - U(2U + 2^{1/2})]\phi_s,$$

with initial conditions

$$\phi_s(-L) = e^{-\alpha L}, \qquad v(-L) = \alpha\phi_s(-L),$$

and integrate it from $z = -L$ to $z = 0$ in order to evaluate $\phi'_s(0)$. Then, by trial and error or by correction (with, say, linear extrapolation), repeat the process until the value of α_s has been found for which $\phi'_s(0)$ is deemed sufficiently small. Show that this gives $\alpha_s = 1.465$ for 'large' values of L.

(iv) Similarly, show that for the varicose mode $\alpha_s = 0.646$.

8.17 *A nonlinear solution for a cat's-eye pattern.* Verify that

$$\psi(x, z) = \log\left[\left(1 + A^2\right)^{1/2}\cosh z + A\cos x\right]$$

is an exact solution of Liouville's equation,

$$\Delta\psi = e^{-2\psi},$$

for all A. Infer that this gives the streamfunction for an exact solution of the vorticity equation of steady two-dimensional flow of an unbounded incompressible inviscid fluid.

Sketch the streamlines for a 'typical' value of A, say $A = 1$. Deduce that

$$\psi(x, z) = \log(\cosh z) + A\,\mathrm{sech}\,z\cos x + O(A^2) \quad \text{as } A \to 0.$$

Hence relate the nonlinear solution above to the neutral stability of the parallel flow with $U(z) = \tanh z$ for $-\infty < z < \infty$.

Also deduce that, in the limit as $A \to \infty$, the solution represents the flow due to an infinite set of equal line vortices of circulation -4π which are spaced at a distance 2π apart on the x-axis and are parallel to the y-axis. [Schmid-Burgk (1965), Stuart (1967, §6), Lamb (1932, p. 224).]

8.18 *Nonlinear wave modes for parallel flows of an incompressible inviscid fluid.* You are given that the vorticity equation of two-dimensional flow of an incompressible inviscid fluid is

$$\frac{\partial\zeta}{\partial t} + \frac{\partial(\psi, \zeta)}{\partial(x, y)} = 0,$$

where ψ is a streamfunction, the vorticity $\zeta = \Delta\psi$ and the velocity components are $u = -\partial\psi/\partial y$, $v = \partial\psi/\partial x$.

Show that if a solution has the form

$$\psi(x, y, t) = F(X, y), \quad \text{where } X = x - ct$$

for some constant c, then f satisfies the nonlinear Poisson equation

$$\Delta F = Q(F + cy), \tag{E8.3}$$

for some function Q of integration.

Show that if $F = \Psi(y)$ is *one* solution of equation (E8.3) and Q is differentiable, then

$$Q'(F + cy) = -K, \tag{E8.4}$$

where K is defined by $K(y) = -U''(y)/[U(y) - c]$.

Show further that if Q is a single-valued function, then either

(a) U is a constant;
(b) U is strictly monotonic and

$$c = U(y_s), \tag{E8.5}$$

where y_s is the unique point such that $U''(y_s) = 0$ within the domain of flow;
(c) equation (E8.5) is satisfied at *all* points y_s where $U''(y_s) = 0$ within the domain of flow; or
(d) c lies outside the range of U within the domain of flow.

Verify that

(a) if $U(y) = \sin y$ for $-\frac{1}{2}\pi < y < \frac{1}{2}\pi$, then $c = 0$;
(b) if $U(y) = 1 - y^2$ for $-1 < y < 1$, then no wave solution exists;
(c) if $U(y) = \operatorname{sech}^2 y$ for $-\infty < y < \infty$, then $c = \frac{2}{3}$; and
(d) if $U(y) = \tanh y$ for $-\infty < y < \infty$, then $c = 0$; unless c lies outside the range of U.

Find the function Q explicitly in each of the above cases for which a wave solution with the given value of c exists.

Next consider the wave solutions f, periodic in X, of equation (E8.3) for the *same* function Q determined by a given parallel flow U. [Barcilon & Drazin (2001)].

8.19 *Cross-flow instability, that is, instability of a spiral flow.* Consider the basic flow of an incompressible inviscid fluid with velocity $\mathbf{U}(z) = U(z)\mathbf{i} + V(z)\mathbf{j}$ bounded by rigid planes at $z = z_1, z_2$, where U, V are given functions. Linearizing Euler's equation of motion, and taking normal modes proportional to $e^{i(\alpha x + \beta y - \alpha ct)}$, show that the stability of the flow is governed by the eigenvalue problem,

$$\left(\tilde{U} - \tilde{c}\right)\left(D^2 \hat{w} - \tilde{\alpha}^2 \hat{w}\right) - \tilde{U}'' \hat{w} = 0,$$

$$\hat{w} = 0 \quad \text{at } z = z_1, z_2,$$

where $D = d/dz$, $\tilde{U} = (\alpha U + \beta V)/\tilde{\alpha}$, $\tilde{c} = \alpha c/\tilde{\alpha}$, for $\tilde{\alpha} = (\alpha^2 + \beta^2)^{1/2} \neq 0$.

Deduce that the stability of the flow is governed by the *infinity* of Rayleigh stability problems for all real values of α, β, so that some three-dimensional perturbation may be unstable when all two-dimensional perturbations are stable. [See Gregory *et al.* (1955).]

8.20 *Rayleigh–Taylor instability of a stratified fluid and internal gravity waves.* You are given that the motion of an incompressible inviscid fluid of variable density under the influence of gravity is governed by Euler's equations, an equation of state, and the equation of continuity, namely,

$$\rho \left(\frac{\partial \mathbf{u}}{\partial t} + \mathbf{u} \cdot \nabla \mathbf{u}\right) = -\nabla p - g\rho\mathbf{k},$$

$$\frac{\partial \rho}{\partial t} + \mathbf{u} \cdot \nabla \rho = 0 \quad \text{and} \quad \nabla \cdot \mathbf{u} = 0,$$

where g is the acceleration due to gravity, $\mathbf{k}$ is the unit vector in the upward direction, and diffusion of density is neglected.

Show that the basic state of rest with

$$\rho(\mathbf{x}, t) = \bar{\rho}(z), \qquad p(\mathbf{x}, t) = \bar{p}(z),$$

gives a solution of the equations of motion if $\bar{p}(z) = p_0 - g \int^z \bar{\rho}(\zeta)\, d\zeta$ for an arbitrary constant p_0 and $\bar{\rho}$ is continuously differentiable.

Taking $\mathbf{u} = (u', v', w')$, $\rho = \bar{\rho} + \rho'$, $p = \bar{p} + p'$, linearizing the equations of motion for small perturbations, and taking normal modes with $w' = \hat{w}(z)e^{i(kx+ly-\omega t)}$ and so forth, show that

$$D(\bar{\rho}D\hat{w}) - \left(k^2 + l^2\right)\bar{\rho}\hat{w} - \frac{g\left(k^2 + l^2\right)}{\omega^2}\frac{d\bar{\rho}}{dz}\hat{w} = 0, \qquad \text{(E8.6)}$$

where $D = d/dz$.

Show further that if the flow is bounded by rigid horizontal plates at $z = z_1, z_2$, then

$$\hat{w}(z) = 0 \quad \text{at } z = z_1, z_2. \qquad \text{(E8.7)}$$

Recognizing that (E8.6), (E8.7) comprise a regular Sturm–Liouville problem, show that there is instability if and only if $d\bar{\rho}/dz > 0$ somewhere in the domain of flow. [Rayleigh (1883) discovered the results of this exercise, so his name was attached to Rayleigh–Taylor instability after Taylor's (1950) work described in §3.7.]

Deduce that if the density is approximated by a constant except in the buoyancy term, in accordance with the Boussinesq approximation, then

$$D^2\hat{w} - \left(k^2 + l^2\right)\hat{w} - \frac{g\left(k^2 + l^2\right)}{\omega^2\bar{\rho}}\frac{d\bar{\rho}}{dz}\hat{w} = 0. \qquad (E8.8)$$

Show that if $\bar{\rho} = \rho_0 e^{-\beta z}$ for $\rho_0, \beta > 0$, and $z_1 = -\infty, z_2 = \infty$, then the eigensolutions are

$$\hat{w}(z) = \text{constant} \times e^{\beta z/2 + imz}, \qquad \omega^2 = \frac{g\beta\left(k^2 + l^2\right)}{k^2 + l^2 + m^2 + \frac{1}{4}\beta^2},$$

for all real k, l, m, representing a continuous spectrum of internal gravity waves. Deduce that in the Boussinesq approximation $\hat{w} = \text{constant} \times e^{imz}$, $\omega = \pm[g\beta(k^2 + l^2)/(k^2 + l^2 + m^2)]^{1/2}$. [Rayleigh (1883, p. 174) observed that 'Contrary to what is met with in most vibrating systems, there is (in the case of stability) a limit (in the case of stability) on the side of rapidity of vibration, but none on the side of slowness.']

8.21 *The instability of a parallel flow of a stratified fluid.* Show that the basic flow

$$\mathbf{u}(x, t) = \mathbf{U}(z), \qquad \rho(\mathbf{x}, t) = \bar{\rho}(z), \qquad p(\mathbf{x}, t) = \bar{p}(z),$$

is a solution of the equations of motion of an incompressible inviscid fluid of variable non-diffusive density under the influence of gravity if

$$\mathbf{U}(z) = U(z)\mathbf{i}, \qquad \bar{p}(z) = p_0 - g\int^z \bar{\rho}(\zeta)\,d\zeta$$

for arbitrary constant p_0 and well-behaved functions $U, \bar{\rho}$.

Considering only two-dimensional motion with

$$\mathbf{u}(\mathbf{x}, t) = U(z)\mathbf{i} + \mathbf{u}'(x, z, t), \qquad \mathbf{u} = u'\mathbf{i} + w'\mathbf{k},$$

$$\rho(\mathbf{x}, t) = \bar{\rho}(z) + \rho'(x, z, t), \qquad p(\mathbf{x}, t) = \bar{p}(z) + p'(x, z, t),$$

linearizing the equations of motion for small perturbations, and taking normal modes of the form

$$w'(x, z, t) = \hat{w}(z)e^{ik(x-ct)},$$

show that

$$(U-c)^2[D(\bar{\rho}D\hat{w})-k^2\bar{\rho}\hat{w}]-(U-c)[D(\bar{\rho}DU)]\hat{w}+\bar{\rho}N^2\hat{w} = 0, \quad \text{(E8.9)}$$

where the *buoyancy frequency* N (sometimes called the *Brunt–Väisälä frequency*) of the basic fluid is defined by $N^2(z) = -g\bar{\rho}'(z)/\bar{\rho}(z)$, and $D = d/dz$.

Show further that if the flow is bounded by rigid horizontal plates at $z = z_1, z_2$, then

$$\hat{w} = 0 \quad \text{at } z = z_1, z_2. \qquad \text{(E8.10)}$$

Deduce that if the density is approximated by a constant except in the buoyancy term, in accordance with the Boussinesq approximation, then

$$(U - c)(\hat{w}'' - k^2\hat{w}) - U''\hat{w} + \frac{N^2}{U - c}\hat{w} = 0. \qquad \text{(E8.11)}$$

This is called the *Taylor–Goldstein equation*. [In spite of equation (E8.11)'s name, it is essentially due independently to Haurwitz (1931) as well as Taylor (1931) and Goldstein (1931).]

8.22 *The Miles–Howard sufficient condition for stability.* Assuming the Taylor–Goldstein problem (E8.11), (E8.10), define

$$H = \hat{w}/(U - c)^{1/2}$$

and show that

$$\int_{z_1}^{z_2}\left[(U - c)(|DH|^2 + k^2|H|^2)\right.$$
$$\left. + \frac{1}{2}U''|H|^2 + \frac{(1/4)U'^2 - N^2}{U - c}|H|^2\right]dz = 0.$$

Taking the imaginary part of this equation, deduce that if $c_i \neq 0$, then

$$\int_{z_1}^{z_2}\left[\left(N^2 - \frac{1}{4}U'^2\right) + k^2|U - c|^2\right]|H|^2/|U - c|^2\, dz = 0,$$

and thence that the condition

$$N^2(z) > \tfrac{1}{4}U'^2(z) \quad \text{for } z_1 \le z \le z_2$$

is sufficient for stability of the basic flow. [Miles (1961), Howard (1961), Drazin & Reid (1981, §44.3).]

8.23 *The Taylor–Goldstein equation for a three-dimensional basic flow.* Show that if a basic flow of a stratified inviscid incompressible fluid has velocity $\mathbf{U}(z) = U(z)\mathbf{i} + V(z)\mathbf{j}$ and density $\bar{\rho}(z)$, then normal modes proportional to $e^{i(kx+ly-kct)}$ are, in the Boussinesq approximation, governed by the equation

$$(kU + lV - kc)\big[D^2\hat{w} - \big(k^2 + l^2\big)\hat{w}\big]$$

$$- \big(kU'' + lV''\big)\hat{w} + \frac{\big(k^2 + l^2\big)N^2}{kU + lV - kc}\hat{w} = 0, \qquad \text{(E8.12)}$$

where $D = d/dz$ and the usual notation is used.

Show that the generalization of the Miles–Howard sufficient condition for stability is that if

$$\big(k^2 + l^2\big)N^2 > \tfrac{1}{4}\big(kU' + lV'\big)^2$$

for $z_1 \le z \le z_2$, then the flow is stable to the *given* mode.

8.24 *The instability of swirling and stratified shear flows.* Consider the stability of an axisymmetric swirling jet of an inviscid incompressible fluid with basic velocity components

$$\mathbf{U}(r) = (0, V(r), W(r))$$

in cylindrical polar coordinates (r, θ, z), and basic pressure $P(r)$, and density $\bar{\rho}(r)$, such that the flow is bounded by rigid cylinders at $r = r_1$, r_2. Linearize the equations of motion for axisymmetric perturbations, take a normal mode equal to the product of $e^{ik(z-ct)}$ and a vector function of r, and deduce that

$$D\big[\bar{\rho}(W - c)^2 D_* F\big] - \bar{\rho}k^2(W - c)^2 F + \Phi F = 0, \qquad \text{(E8.13)}$$

where $D = d/dr$, $D_* = r^{-1}Dr$, $\Phi = r^{-3}D(\bar{\rho}r^2 V^2)$, $F = \hat{u}/ik(W - c)$ and $u'_r = \hat{u}(r)e^{ik(z-ct)}$ is the perturbation of the r-component of the velocity.

*Comment on the analogy between equation (E8.13) and the Taylor–Goldstein equation (E8.9). Deduce that the swirling flow is stable if

$$\Phi/(DW)^2 > \tfrac{1}{4}$$

throughout the domain of flow. [Leibovich (1969, 1979).]

8.25 *A three-dimensional instability of unbounded plane Couette flow of a viscous fluid.* Suppose that $u(\mathbf{x}, t) = \eta(t)z - \zeta(t)y$, $v(\mathbf{x}, t) = -\tfrac{1}{2}\xi(t)z$,

$w(\mathbf{x}, t) = \frac{1}{2}\xi(t)y$ for some functions ξ, η, ζ. Show that $\nabla \cdot \mathbf{u} = 0$ and the vorticity $\boldsymbol{\omega} = (\xi, \eta, \zeta)$. Deduce that this vorticity field is an exact solution of the vorticity equation for a viscous incompressible fluid provided only that

$$\frac{d\xi}{dt} = 0, \qquad \frac{d\eta}{dt} = -\frac{1}{2}\zeta\xi, \qquad \frac{d\zeta}{dt} = \frac{1}{2}\xi\eta.$$

Why does the viscosity not affect the dynamics?

Defining the perturbations $\xi' = \xi$, $\eta' = \eta - \sigma$, $\zeta' = \zeta$ for some constant basic shear σ, and linearizing, show that

$$\frac{d\xi'}{dt} = 0, \qquad \frac{d\eta'}{dt} = 0, \qquad \frac{d\zeta'}{dt} = \frac{1}{2}\sigma\xi'.$$

Deduce that $\xi'(t) = \xi'(0)$, $\eta'(t) = \eta'(0)$, $\zeta'(t) = \zeta'(0) + \frac{1}{2}\sigma\xi'(0)t$, so that the perturbation grows linearly with t, and the basic flow $\mathbf{U} = \sigma z \mathbf{i}$ appears to be unstable (although the initial velocity perturbation is not bounded in space).

Further, show without linearization that $\xi(t) = \xi(0)$, $\eta(t) = \eta(0)\cos[\frac{1}{2}\xi(0)t] + \zeta(0)\sin[\frac{1}{2}\xi(0)t]$, and $\zeta(t) = \zeta(0)\cos[\frac{1}{2}\xi(0)t] - \eta(0)\sin[\frac{1}{2}\xi(0)t]$, and that therefore the basic state $\xi = 0, \eta = \sigma, \zeta = 0$ is indeed unstable, although the perturbation is bounded for all time. [Waleffe (1995). See Exercise 8.8.]

8.26 *The energy equation.* Show that if $\mathbf{U} = U(z)\mathbf{i}$ for $z_1 \leq z \leq z_2$, then the Reynolds–Orr equation (5.28) for a two-dimensional perturbation of period $2\pi/\alpha$ in x becomes

$$\frac{dK}{dt} = M - R^{-1}N$$

in dimensionless variables, where

$$K = \int_{z_1}^{z_2} \int_0^{2\pi/\alpha} \frac{1}{2}\left(u'^2 + w'^2\right) dx\, dz,$$

$$M = -\int_{z_1}^{z_2} \int_0^{2\pi/\alpha} \frac{dU}{dz} u'w'\, dx\, dz,$$

$$N = \int_{z_1}^{z_2} \int_0^{2\pi/\alpha} \left(\frac{\partial u'}{\partial z} - \frac{\partial w'}{\partial x}\right)^2 dx\, dz.$$

Hence deduce equation (8.54). Interpret the physical meanings of K, M and N.

8.27 *Stability problem of the asymptotic-suction boundary layer.* Verify that the vorticity equation,

$$\frac{\partial(\Delta\psi)}{\partial t} + \frac{\partial(\Delta\psi, \psi)}{\partial(x, z)} = \nu\Delta^2\psi,$$

for two-dimensional flow of an incompressible viscous fluid is satisfied by $\psi = \Psi$ for $z \geq 0$, where

$$\Psi(x, z) = W_0 x + U_0[z + (\nu/W_0)\exp(-W_0 z/\nu)]$$

for arbitrary U_0, W_0, ν. Deduce that this gives the basic flow $\mathbf{U} = (U, V, W)$, where

$$U(z) = U_0[1 - \exp(-W_0 z/\nu)], \qquad V = 0, \qquad W(z) = -W_0.$$

Taking $\psi = \Psi + \psi'$, linearizing, and assuming a normal mode of the form $\psi'(x, z, t) = \phi(z)\exp[i\alpha(x - ct)]$, show that

$$\nu\big(\phi^{\mathrm{iv}} - 2\alpha^2\phi'' + \alpha^4\phi\big) = i\alpha\big[(U-c)\big(\phi'' - \alpha^2\phi\big) - U''\phi\big] - W_0\big(\phi''' - \alpha^2\phi'\big),$$

where

$$\phi(z) = \phi'(z) = 0 \quad \text{at } z = 0, \qquad \phi(z), \phi'(z) \to 0 \quad \text{as } z \to \infty.$$

[Hughes & Reid (1965). In fact, $c = 0.150$ for $R_c = 54\,370$, $a_c = 0.1555$, on taking length scale $L = \nu/W_0$ so that the Reynolds number is defined as $R = U_0/W_0$ (Hocking, 1975).]

8.28 *Squire's transformation.* First note that, given an unstable parallel flow, Squire's transformation shows that the relative growth rate of a mode with wavenumbers (α, β) is $\sigma(\alpha, \beta, R) = \alpha c_i(\tilde{\alpha}, \alpha R/\tilde{\alpha})$, where $\tilde{\alpha} = (\alpha^2 + \beta^2)^{1/2}$.

Defining $\sigma_2(R) = \max_{\alpha\geq 0} \sigma(\alpha, 0, R)$, $\sigma_3(R) = \max_{\alpha,\beta\geq 0} \sigma(\alpha, \beta, R)$, and $\sigma_M = \max_{\alpha,\beta,R\geq 0} \sigma(\alpha, \beta, R)$, and assuming that $\max_{R\geq 0} \sigma_2(R) = \sigma(R_M)$, show that $\sigma_M = \sigma_2(R_M)$, the maximum being attained for some two-dimensional mode of wavenumber α_M, say, at Reynolds number $R_M > R_c$. *Assuming* that $\sigma_2(R)$ increases monotonically from zero to σ_M as R increases from R_c to R_M, show that if $R_c < R \leq R_M$, then $\sigma_2(R) > \sigma_3(R)$, that is, the fastest-growing mode is two-dimensional, but if $R > R_M$, then the fastest-growing mode *might* be three-dimensional.

[Watson (1960a). Of all the calculations made, none suggests that there is some value of R for which the fastest-growing mode for some basic flow is three-dimensional. In any event, R_M is usually so large that the flow is turbulent, and the theoretical possibility that the fastest-growing linear mode is three-dimensional has no practical importance. In fact, for plane Poiseuille flow $\sigma_M \approx 0.6$ at $\alpha_M \approx 0.8$, $R_M \approx 50\,000$.]

8.29 *A simple exact solution of the Orr–Sommerfeld equation.* Show that the Navier–Stokes equations are invariant under the continuous group of translations $z \mapsto z + \delta$ for all real δ. Deduce that a continuously twice-differentiable basic velocity $U(z)\mathbf{i} \mapsto U(z + \delta)\mathbf{i} = U(z)\mathbf{i} + \delta U'(z)\mathbf{i} + O(\delta^2)$ as $\delta \to 0$. Hence or otherwise show that a solution of the Orr–Sommerfeld equation is given by $\alpha = \beta = 0$, $\phi = U$ for all c, R, provided that the basic flow is an exact solution of the Navier–Stokes equations. Does this give an eigensolution?

8.30 *Another simple exact solution of the Orr–Sommerfeld equation, for uniform flow.* Show that if $U = $ constant, then the general solution of the Orr–Sommerfeld equation is

$$\phi(z) = A_1 e^{\alpha z} + A_2 e^{-\alpha z} + A_3 e^{\gamma z} + A_4 e^{-\gamma z},$$

where A_1, A_2, A_3, A_4 are arbitrary constants and $\gamma^2 = \alpha^2 + i\alpha R(U - c)$. [In fact the solution $\psi' = e^{\pm \gamma z - i\alpha ct}$ when $\alpha = 0$, $\alpha c \neq 0$ corresponds to Stokes waves due to an oscillating plane wall (see, e.g., Batchelor, 1967, §4.3).]

8.31 *Stability of a uniform flow of an incompressible viscous fluid.* Suppose that an incompressible fluid of kinematic viscosity ν has basic velocity $\mathbf{U}_* = V\mathbf{i}$ and is bounded by rigid planes (moving with the basic velocity $V\mathbf{i}$) at $z_* = \pm L$. Deduce that two-dimensional normal modes proportional to $e^{i\alpha(x - ct)}$ are governed by dimensionless eigenvalue relations

$$c = 1 - i(\alpha^2 + p_n^2)/\alpha R, \ 1 - i(\alpha^2 + q_n^2)/\alpha R,$$

where $R = VL/\nu$ and $p_n(\alpha)$ are the positive roots of

$$p \tan p = -\alpha \tanh \alpha,$$

for $n = 0, 2, 4, \ldots$, and $q_n(\alpha)$ of

$$q \cot q = \alpha \coth \alpha,$$

for $n = 1, 3, 5, \ldots$. Deduce that the flow is stable. [Rayleigh (1892).]

8.32 *Schensted modes.* Assuming that the eigensolutions of the Orr–Sommerfeld problem for $-1 \leq z \leq 1$ can be expressed as

$$\phi(z) = \sum_{j=0}^{\infty} (i\alpha R)^j \phi^{(j)}(z), \qquad c = (i\alpha R)^{-1} \sum_{j=0}^{\infty} (i\alpha R)^j c^{(j)}$$

for sufficiently small αR and fixed α, and that term-by-term differentiation is justified, show that

$$\left(D^2 - \alpha^2 + c^{(0)}\right)\left(D^2 - \alpha^2\right)\phi^{(0)} = 0$$

and

$$\phi^{(0)}(z) = D\phi^{(0)}(z) = 0 \quad \text{at } z = \pm 1.$$

Using the result of Example 8.6 for sinuous modes, assume that

$$c^{(0)} = \alpha^2 + p_n^2, \qquad \phi^{(0)}(z) = \frac{\cosh \alpha z}{\cosh \alpha} - \frac{\cos p_n z}{\cos p_n},$$

for $n = 0, 2, 4, \ldots$, where $p_n(\alpha)$ is the $\frac{1}{2}(n + 2)$th positive root of

$$\alpha \tanh \alpha + p \tan p = 0.$$

Show that $p_n(0) = \frac{1}{2}(n+2)\pi$ and thence find $c^{(0)}$ and $\phi^{(0)}$ when $\alpha = 0$. Show that

$$\left(D^2 - \alpha^2 + c^{(0)}\right)\left(D^2 - \alpha^2\right)\phi^{(1)} = \left[(U - c^{(1)})(D^2 - \alpha^2) - U''\right]\phi^{(0)},$$

$$\phi^{(1)}(z) = D\phi^{(1)}(z) = 0 \quad \text{at } z = \pm 1.$$

Without solving this problem for $\phi^{(1)}$, deduce the solvability condition

$$\int_{-1}^{1} \phi^{(0)}(z)\left[(U - c^{(1)})(D^2 - \alpha^2) - U''\right]\phi^{(0)} \, dz = 0$$

to determine $c^{(1)}$.

Show that if $U(z) = 1 - z^2$ and $\alpha = 0$, then

$$c^{(1)} = \frac{2}{3} - \frac{5}{2p_n^2}$$

for $n = 0, 2, 4, \ldots$.

[Pekeris (1936). Hint: $\int_{-1}^{1} \cos p_n z \, dz = 0$, $\int_{-1}^{1} \cos^2 p_n z \, dz = 1$, $\int_{-1}^{1} z^2 \cos p_n z \, dz = 4 \cos p_n / p_n^2$, $\int_{-1}^{1} z^2 \cos^2 p_n z \, dz = \frac{1}{3} + 1/2p_n^2$.]

8.33 *A marginal curve.* Suppose that the dispersion relation for a normal mode proportional to $e^{i\alpha x + st}$ is of the form $\mathcal{F}(\alpha, s, R) = 0$ in some 'toy' problem, where $R > 0$ is a 'Reynolds number', and

$$\mathcal{F}(\alpha, s, R) = Rs + \left(\alpha^8 R^2 - \alpha^3 R + 1\right) + i\alpha R.$$

Then compute or sketch the marginal curve in the (R, α)-plane, and show that $R_c = \frac{4096}{225}(\frac{3}{5})^{1/2} \approx 14.1$.

8.34 *Gaster's transformation.* Suppose that the dispersion relation for a normal mode proportional to $e^{i\alpha x + st}$ is of the form

$$\mathcal{F}(\alpha, s, R) = 0,$$

where R is a Reynolds number, and $\mathcal{F}$ a complex dispersion function. Then show that, for a temporal mode, the neutral curve in the (R, α)-plane is defined by the roots of the pair of real equations,

$$\mathrm{Re}[\mathcal{F}(\alpha, -i\omega, R)] = 0, \qquad \mathrm{Im}\,[\mathcal{F}(\alpha, -i\omega, R)] = 0,$$

on regarding $\omega = is$ as a real parameter of the curve. Show that, for a spatial mode, the neutral curve in the (R, ω)-plane is defined by the same pair of equations, on regarding α as a real parameter of the curve.

Taking $s = \sigma - i\omega$, $\alpha = \alpha_r + i\alpha_i$, suppose that s is an analytic function of α, use a Cauchy–Riemann relation, and deduce that for a spatial mode

$$\frac{\partial\sigma}{\partial\alpha_i} = -c_g, \qquad\qquad (E8.14)$$

where the group velocity $c_g = \partial\omega/\partial\alpha_r$. [This formula is useful, on evaluating $\partial\sigma/\partial\alpha_i$ where $\alpha_i = 0$, to find the relative growth rate α_i of spatial modes in terms of the temporal relative growth rate σ near a marginal curve, by evaluating the group (*not* the phase) velocity, and hence to determine on which side of the curve instability lies.]

Suppose that for a particular model problem

$$\frac{\partial u}{\partial t} + U\frac{\partial u}{\partial x} - u + a\frac{\partial^3 u}{\partial x^3} = R^{-1}\frac{\partial^2 u}{\partial x^2},$$

where $a > 0$. Deduce that the dispersion function is

$$\mathcal{F}(\alpha, s, R) = s - 1 + \alpha^2/R + i\alpha U - ia\alpha^3.$$

Then sketch the neutral curves in the (R, α)- and (R, ω)-planes, and verify Gaster's transformation (E8.14). [Gaster (1962).]

8.35 *Decay of spatial modes.* Show that the growth or decay in space of steady perturbations of a basic parallel flow of a viscous incompressible fluid with velocity $U(z)\mathbf{i}$ for $-1 \leq z \leq 1$ is governed by the eigenvalue problem

$$\phi^{\text{iv}} - 2\alpha^2\phi'' + \alpha^4\phi = \mathrm{i}\alpha R\left[U\left(\phi'' - \alpha^2\phi\right) - U''\phi\right],$$

where $\phi = \phi' = 0$ at $z = \pm 1$.

Hence show that if $R = 0$, then either

$$\phi(z) = \frac{\cos\alpha z}{\cos\alpha} - \frac{z\sin\alpha z}{\sin\alpha}, \qquad \sin 2\alpha = -2\alpha,$$

or

$$\phi(z) = \frac{\sin\alpha z}{\sin\alpha} - \frac{z\cos\alpha z}{\cos\alpha}, \qquad \sin 2\alpha = 2\alpha.$$

Deduce that $\alpha = \alpha_n, \alpha_n^*$ for $n = 0, 1, 2, \ldots$, where

$$\alpha_n \sim \tfrac{1}{4}(2n+3)\pi + \tfrac{1}{2}\mathrm{i}\log[(2n+3)\pi] \quad \text{as } n \to \infty.$$

[Wilson (1969). In fact, $\alpha_0 \approx 2.10620 + 1.12536\mathrm{i}$, $\alpha_1 \approx 3.74884 + 1.38434\mathrm{i}$, $\alpha_2 \approx 5.35627 + 1.55158\mathrm{i}$.]

8.36 *A less simple exact solution of the Orr–Sommerfeld equation, for plane Couette flow.* Show that if $U(z) = z$, then the general solution of the Orr–Sommerfeld equation is

$$\phi(z) = A_1\mathrm{e}^{\alpha z} + A_2\mathrm{e}^{-\alpha z} + \frac{1}{\alpha}\int^z \sinh[k(y-z)]\Omega(y)\,\mathrm{d}y,$$

where

$$\Omega(y) = A_3\mathrm{Ai}(w) + A_4\mathrm{Bi}(w),$$

A_1, A_2, A_3, A_4 are arbitrary constants, Ai, Bi are Airy functions, and $w = \mathrm{e}^{\mathrm{i}\pi/6}(\alpha R)^{1/3}(y - c - \mathrm{i}\alpha/R)$.

8.37 *The stabilizing effect of boundaries.* Is plane Couette flow of an incompressible viscous fluid stable or unstable? For what values of the Reynolds number?

Consider a given smooth parallel flow U of a viscous incompressible fluid confined by rigid planes at $z = z_1, z_2$, and the limit as $z_2 \to z_1$ for fixed U and Reynolds number; then what is the name of the well-known flow that approximates U asymptotically within the domain of flow? Deduce heuristically that *all* smooth flows are stable in the above limit. [Cf. Exercise 8.15.]

8.38 *The stability to short waves.* Consider this argument. 'A small localized
perturbation of a basic shear layer composed of short-wave components
is sensitive only to the local properties of the basic flow, because the influence of the perturbation decays exponentially and rapidly with distance
from the locality of the perturbation. Therefore the perturbation is locally
stable, as the plane Couette flow approximated locally by the basic flow
is.' If you believe this crude argument, substantiate it as well as you can;
if you disbelieve it, refute it.

8.39 *Boundary conditions for the Orr–Sommerfeld problem when the basic
velocity profile is discontinuous.* Show that if U or U' is discontinuous at
$z = z_0$, then a solution ϕ of the Orr–Sommerfeld equation satisfies the
'jump' boundary conditions

$$\Delta[\phi] = 0, \qquad \Delta[D\phi] = 0, \qquad \Delta\big[D^2\phi + i\alpha R(U - c)\phi\big] = 0,$$

$$\Delta\big[D^3\phi - i\alpha R\{(U - c)D\phi - U'\phi\}\big] = 0$$

at $z = z_0$, where $D = d/dz$. [Here Δ denotes the jump of the contents
of the brackets that follow, not the Laplacian operator.]

[Hint: Integrate the Orr–Sommerfeld equation across the discontinuity
four times. It can be shown that although a discontinuous velocity profile
is incompatible with the equations of motion of a viscous fluid, the use
of a discontinuous profile in solving the Orr–Sommerfeld equation may
be justified as giving an approximate solution for long-wave modes of a
jet or shear layer. Drazin (1961).]

Hence show that if you take the basic vortex sheet with

$$U(z) = \begin{cases} 1 & \text{for } z > 0, \\ -1 & \text{for } z < 0, \end{cases}$$

then

$$\Delta[\phi] = 0, \quad \Delta[D\phi] = 0, \quad \Delta\big[(D^2+\beta^2)\phi\big] = 0, \quad \Delta\big[(D^2-\beta^2)D\phi\big] = 0$$

at $z = 0$, where $\beta^2 = \alpha^2 + i\alpha R(U - c)$. Deduce that

$$c_r = 0, \qquad R\big(1 - 3^{1/2}c_i\big)^2 = 4\alpha\big(3^{1/2} - c_i\big),$$

and thence that the marginal curve has equation

$$R = 4 \times 3^{1/2}\alpha,$$

and that

$$c \sim -4i\alpha/3R \quad \text{as } R/\alpha \to 0.$$

[Tatsumi & Gotoh (1960).]

8.40 *The adjoint Orr–Sommerfeld problem.* Define the linear operator L_c by

$$\mathrm{L}_c\phi = \left(\mathrm{D}^2 - \alpha^2\right)^2\phi - i\alpha R\big[(U - c)\left(\mathrm{D}^2 - \alpha^2\right)\phi - U''\phi\big]$$

for complex c and for all $\phi \in S$, for given $U, R > 0, z_1 < z_2, \alpha > 0$, where $\mathrm{D} = d/dz$ as usual, and where S is the set of all complex-valued functions over the interval $[z_1, z_2]$ with continuous fourth derivatives. Deduce, by integration by parts, that

$$\int_{z_1}^{z_2} (\mathrm{L}_c\phi)\psi^* \, dz = \int_{z_1}^{z_2} \phi\left(\mathrm{L}_c^\dagger\psi^*\right) dz + \big[\psi^*\phi''' - \psi^{*'}\phi'' + \psi^{*''}\phi'$$
$$- \psi^{*'''}\phi - 2\alpha^2\left(\psi^*\phi' - \psi^{*'}\phi\right)$$
$$- i\alpha R\big\{(U - c)(\phi'\psi^* - \phi\psi^{*'}) - U'\phi\psi^*\big\}\big]_{z_1}^{z_2}$$

for all $\phi, \psi \in S$, where the adjoint operator $\mathrm{L}_c^\dagger$ is defined by

$$\mathrm{L}_c^\dagger\psi^* = \left(\mathrm{D}^2 - \alpha^2\right)^2\psi^* + i\alpha R\big[\left(\mathrm{D}^2 - \alpha^2\right)(U - c^*)\psi^* - U''\psi^*\big].$$

Suppose next that

$$\mathrm{L}_c\phi = 0 \quad \text{and} \quad \phi(z) = \mathrm{D}\phi(z) = 0 \quad \text{at } z = z_1, z_2$$

determines an eigensolution c, ϕ of the Orr–Sommerfeld problem, and

$$\mathrm{L}_d^\dagger\psi^* = 0, \quad \text{and} \quad \psi^*(z) = \mathrm{D}\psi^*(z) = 0 \quad \text{at } z = z_1, z_2$$

determines an eigensolution d, ψ of the adjoint Orr–Sommerfeld problem. Deduce that either $d = c^*$ or

$$\int_{z_1}^{z_2} \phi\left(\mathrm{D}^2 - \alpha^2\right)\psi^* \, dz = 0.$$

[Note that

$$\left(\mathrm{D}^2 - \alpha^2\right)^2\psi = i\alpha R\big[\left(\mathrm{D}^2 - \alpha^2\right)(U - d)\psi - U''\psi\big],$$

$$\psi(z) = \mathrm{D}\psi(z) = 0 \quad \text{at } z = z_1, z_2,$$

and that

$$\int_{z_1}^{z_2} \phi\left(\mathrm{D}^2 - \alpha^2\right)\psi^* \, dz = -\int_{z_1}^{z_2} \big[(\mathrm{D}\phi)(\mathrm{D}\psi^*) + \alpha^2\phi\psi^*\big] dz,$$

on integration by parts and use of the boundary conditions. It is found that in fact the set $\{d\}$ of eigenvalues of $L_d^{\dagger}$ above is the complex conjugate of the set $\{c\}$ of eigenvalues of L_c, although the eigenfunctions are not simply related. Hint: How is this exercise related to Exercise 5.9?]

8.41 *A numerical method of solution of the Orr–Sommerfeld problem for spatial modes.* Consider the Orr–Sommerfeld problem for a spatial mode, namely,

$$\phi^{\text{iv}} - 2\alpha^2\phi'' + \alpha^4\phi = iR\big[(\alpha U - \omega)(\phi'' - \alpha^2\phi) - \alpha U''\phi\big],$$

$$\phi(z) = \phi'(z) = 0 \quad \text{at } z = z_1, z_2,$$

where $\omega = \alpha c$ is a given real frequency of the mode.

Show that this problem is equivalent to

$$\frac{\mathrm{d}}{\mathrm{d}z}\begin{bmatrix} u_1 \\ u_2 \\ u_3 \\ u_4 \end{bmatrix} = \begin{bmatrix} u_2 \\ \alpha^2 u_1 + u_3 \\ u_4 \\ -i\alpha R U'' u_1 + iR(\alpha U - \omega)u_3 + \alpha^2 u_3 \end{bmatrix},$$

[Hint: Let $u_1 = \phi$, $u_3 = \phi'' - \alpha^2\phi$.]

You are given that, by some spectral expansion of the eigenfunction, this problem is reduced approximately to the algebraic eigenvalue problem,

$$\mathbf{A}\mathbf{x} = \alpha\mathbf{B}\mathbf{x} + \alpha^2\mathbf{C}\mathbf{x},$$

where $\mathbf{A}, \mathbf{B}, \mathbf{C}$ are some $n \times n$ matrices (dependent on U, U'', ω and R) and $\mathbf{x}$ is an n-column vector (representing the eigenfunction u_1). Defining

$$\mathbf{y} = \alpha\mathbf{x}$$

and the $2n$-column vector

$$\mathbf{z} = \begin{bmatrix} \mathbf{x} \\ \mathbf{y} \end{bmatrix},$$

show that the algebraic problem implies the block matrix equations

$$\begin{bmatrix} \mathbf{A} & \mathbf{0} \\ \mathbf{0} & \mathbf{I} \end{bmatrix}\mathbf{z} = \alpha\begin{bmatrix} \mathbf{B} & \mathbf{C} \\ \mathbf{I} & \mathbf{0} \end{bmatrix}\mathbf{z}$$

and

$$\begin{bmatrix} \mathbf{A} & -\mathbf{B} \\ \mathbf{0} & \mathbf{I} \end{bmatrix}\mathbf{z} = \alpha\begin{bmatrix} \mathbf{0} & \mathbf{C} \\ \mathbf{I} & \mathbf{0} \end{bmatrix}\mathbf{z}.$$

8.42 *The Orr–Sommerfeld problem with a cross flow.* Taking a basic velocity $\mathbf{U} = U(z)\mathbf{i}$ and normal modes of the form

$$w'(\mathbf{x}, t) = \hat{w}(z)e^{i\boldsymbol{\alpha}\cdot\mathbf{x}-\omega t}$$

and so forth, where $\boldsymbol{\alpha} = \alpha\mathbf{i} + \beta\mathbf{j}$, show that the Orr–Sommerfeld equation can be re-written as

$$\left(D^2 - \alpha^2\right)^2\hat{w} = iR\left[(\boldsymbol{\alpha}\cdot\mathbf{U} - \omega)\left(D^2 - \alpha^2\right)\hat{w} - \boldsymbol{\alpha}\cdot\left(d^2U/dz^2\right)\hat{w}\right].$$

Hence or otherwise show that if the basic velocity is $\mathbf{U} = U(z)\mathbf{i} + V(z)\mathbf{j}$, then the linearized equation again has the above form.

8.43 *A derivation of the Squire and Orr–Sommerfeld problems.* Consider a basic flow of a viscous incompressible fluid with velocity $\mathbf{U} = U(z)\mathbf{i}$ between fixed rigid planes at $z = z_1, z_2$. Taking $\mathbf{u} = \mathbf{U} + \mathbf{u}'$, $p = P + p'$ in the usual way, linearize the Navier–Stokes equations and deduce that

$$\left(\frac{\partial}{\partial t} + U\frac{\partial}{\partial x}\right)\Delta w' - U''\frac{\partial w'}{\partial x} - R^{-1}\Delta^2 w' = 0, \quad \text{(E8.15)}$$

$$\frac{\partial\zeta'}{\partial t} + U\frac{\partial\zeta'}{\partial x} - R^{-1}\Delta\zeta' = U'\frac{\partial w'}{\partial y}, \quad \text{(E8.16)}$$

where $U'' = d^2U/dz^2$ and the z-component of the vorticity of the perturbation is $\zeta' = \partial v'/\partial x - \partial u'/\partial y$.

Show that the solution is a superposition of

(a) modes which are proportional to $e^{i(\alpha x+\beta y-\alpha ct)}$, for which $c = c(\tilde{\alpha}^2, \alpha R)$ is an eigenvalue of the Orr–Sommerfeld problem

$$\left(D^2 - \tilde{\alpha}^2\right)^2 w' - i\alpha R\left[(U - c)\left(D^2 - \tilde{\alpha}^2\right) - U''\right]w' = 0,$$

$$w'(\mathbf{x}, t) = Dw'(\mathbf{x}, t) = 0 \quad \text{at } z = z_1, z_2,$$

where $D = d/dz$; and for which

$$\left[D^2 - \tilde{\alpha}^2 - i\alpha R(U - c)\right]\zeta' = i\beta U'w',$$

$$\zeta'(\mathbf{x}, t) = 0 \quad \text{at } z = z_1, z_2;$$

and of

(b) modes which are proportional to $e^{i(\alpha x+\beta y-\alpha dt)}$, for which $d = d(\tilde{\alpha}^2, \alpha R)$ is an eigenvalue of the homogeneous Squire problem

$$\left[D^2 - \tilde{\alpha}^2 - i\alpha R(U - d)\right]\zeta' = 0,$$

$$\zeta'(\mathbf{x}, t) = 0 \quad \text{at } z = z_1, z_2;$$

Show that if $\zeta' = \hat{\zeta}(z)e^{i(\alpha x + \beta y - \alpha dt)}$ for a Squire mode, then

$$\int_{z_1}^{z_2} \left[|D\hat{\zeta}|^2 + \left(\tilde{\alpha}^2 + \alpha R d_i\right)|\hat{\zeta}|^2\right] dz = 0.$$

Deduce that all the Squire modes are stable.

8.44 *Stable equilibrium, yet with large perturbations.* Find the general solution of the system

$$\frac{dA}{dt} = -cA, \quad \frac{dB}{dt} = A - dB, \quad A(0) = A_0, \quad B(0) = B_0 \neq 0.$$

(i) Suppose further that $c = 2R^{-1}, d = R^{-1}$ for $R > 0$, and thence show that $\max_{t \geq 0} |B(t)| = (RA_0 + B_0)^2/4R|A_0| \sim \frac{1}{4}R|A_0|$ as $R \to \infty$ for fixed A_0, B_0. Sketch the phase portrait.

[This models crudely the linear interaction of an Orr–Sommerfeld mode of amplitude A and a Squire mode of amplitude B. Note how an initial perturbation of an asymptotically stable point of equilibrium may be amplified very strongly when R is large. See Exercise 2.8 and Trefethen *et al.* (1993).]

(ii) Next solve the initial-value problem for the differential system in the special case when $d = c$. What is $\max_{t \geq 0} |B(t)|$ when $c = R^{-1}$ and R is large?

[This models the exceptional case of resonance between stable Squire and Orr–Sommerfeld modes.]

8.45 *Mothers and daughters.* Consider this simple linear ordinary-differential system as merely a 'toy' model of the evolution of two modes of amplitudes A_1, A_2:

$$\frac{d\mathbf{A}}{dt} = \mathbf{L}\mathbf{A},$$

where

$$\mathbf{A} = \begin{bmatrix} A_1 \\ A_2 \end{bmatrix}, \quad \mathbf{L} = \begin{bmatrix} i\omega - \sigma + \epsilon & in \\ 0 & i\omega - \sigma \end{bmatrix},$$

for given real constants $\sigma > 0, \epsilon > 0, \omega, n$. Show that the general solution of this system can be written in the form

$$\mathbf{A}(t) = \alpha \begin{bmatrix} 1 \\ 0 \end{bmatrix} e^{(-\sigma + \epsilon + i\omega)t} + \beta \begin{bmatrix} 1 \\ i\epsilon/n \end{bmatrix} e^{(-\sigma + i\omega)t}$$

for some complex constants α, β.

Define the vectors

$$\mathbf{d} = \begin{bmatrix} 1 \\ 0 \end{bmatrix}, \quad \mathbf{m} = \begin{bmatrix} 0 \\ 1 \end{bmatrix},$$

and whimsically call them daughter and mother respectively. Then consider the initial-value problems for which $\mathbf{A}(0) = \mathbf{d}$ and $\mathbf{A}(0) = \mathbf{m}$. Show that the respective solutions of these two problems are

$$\mathbf{A}(t) = \mathbf{d}e^{(-\sigma+\epsilon+i\omega)t}$$

and

$$\mathbf{A}(t) = i n \epsilon^{-1}\mathbf{d}(e^{\epsilon t} - 1)e^{(-\sigma+i\omega)t} + \mathbf{m}e^{(-\sigma+i\omega)t},$$

$$\rightarrow int\mathbf{d}e^{(-\sigma+i\omega)t} + \mathbf{m}e^{(-\sigma+i\omega)t} \quad \text{as } \epsilon \rightarrow 0 \text{ for fixed } t.$$

On the basis of these results, explain Boberg & Brosa's (1988) summary: (1) a motherless daughter fades; (2) a mother must produce a daughter; (3) a daughter grows as long as her mother lives.

*Apply this analogy to the large transient growth of modes in a stable flow such as plane Couette flow or Poiseuille pipe flow. [Boberg & Brosa (1988).]

9

Routes to Chaos and Turbulence

Chaos is come again.
Othello III *3, line 92*

In this chapter, we shall draw together some general features of the onset of chaos and turbulence. The theory of dynamical systems, and in particular the theories of bifurcation and chaos, provide a mathematical framework with which we may interpret *qualitatively* the transition to turbulence without having to clutter our minds with a lot of detail. This framework can be used together with physical arguments of the mechanics of transition to understand the essence of instability of flows which may be so complicated geometrically as to defy solution except in numerical terms. However, the dynamics of fluids is very diverse, and the details of transition to turbulence depend on the details of the flow undergoing transition, and therefore can only be found by careful experiments and computational fluid dynamics of each case.

9.1 Evolution of Flows as the Reynolds Number Increases

The details of transition to turbulence not only are complicated but also vary greatly from flow to flow, so there is no possibility of a short summary of all transition. However, there are some unifying themes in the theory, and a few routes to turbulence essentially shared by many flows, even though the physical mechanisms of the same route may differ from one flow to another sharing the same route.

One major theme is that, as the Reynolds number (or Rayleigh number, Taylor number or other parameter measuring the speed of a class of dynamically similar flows with a given steady configuration) increases, the temporal and spatial complexity of observed flows often increases in a succession of bifurcations until the onset of turbulence. Each bifurcation is marked by the onset of instability of one flow and followed by equilibration to another stable flow, steady or unsteady. This idea goes back to the theory of Landau (1944) and the experiments of Malkus (1954), who interpreted in this way his measurements of nonlinear Rayleigh–Bénard convection as the Rayleigh number increased. The idea has been developed and refined by many others subsequently.

208

Let us then trace the evolution of stable flows in a given steady bounded configuration as the Reynolds number R increases quasi-statically. When $R = 0$ there is Stokes flow, which is steady and unique. Serrin's theorem shows that when R is sufficiently small there is stable steady flow and it is the unique steady flow. As R increases, other steady flows may arise. Benjamin (1976) has shown (by use of Leray–Schauder degree theory) that there is in general (effectively for all values of R except those where there is a bifurcation) an *odd* number of steady flows. Thus turning points, transcritical bifurcations and pitchfork bifurcations may arise to give multiple steady solutions at some values of R. Some, all or none of these steady solutions are stable. For example, Benjamin & Mullin (1982) observed 20 distinct stable steady sets of Taylor vortices in one fairly short pair of rotating cylinders in different experiments with the same angular velocities, and inferred the coexistence of 19 unstable steady solutions – the set of steady stable Taylor vortices observed in a given experiment depends upon how the experiment is set up, that is, what the initial conditions are and how the angular velocities of the cylinders are made to reach their prescribed values.

If, as R increases, a steady solution becomes unstable where $\mathrm{Re}(s) = 0$ and $\mathrm{Im}(s) = 0$, then we expect a turning point in general. Exceptionally a transcritical bifurcation or a pitchfork bifurcation may occur; if the flow has a reflectional symmetry, then a pitchfork bifurcation is to be expected. If, however, $\mathrm{Re}(s) = 0$ and $\mathrm{Im}(s) \neq 0$ where a steady solution loses stability, then we expect a Hopf bifurcation and the branching of either a supercritical stable time-periodic solution for $R > R_c$ or a subcritical unstable time-periodic solution for $R < R_c$. An example of a supercritical Hopf bifurcation occurs when the wake behind a bluff body becomes unstable and a vortex street forms; another occurs when Taylor vortices become wavy. This shows the practical importance of the validity or invalidity of the principle of exchange of stabilities, although that principle may seem unimportant in the context of the linear theory alone. The vanishing or not of $\mathrm{Im}(s)$ at marginal stability determines whether the bifurcation is to a steady or time-periodic flow; weak nonlinearity at marginal stability determines further the type of bifurcation, whether it is supercritical or subcritical, and so forth.

After a supercritical Hopf bifurcation, the stable time-periodic flow evolves as R increases further, and may itself eventually become unstable. The instability may be found mathematically by linearization for small perturbations of the time-periodic flow and use of Floquet theory, which is to the instability of periodic solutions of differential equations as the method of normal modes is to the stability of steady solutions. Periodic solutions of differential equations are subject to *parametric instability*; as in the method of normal modes, parametrically unstable modes may have a new period of any value. But a common type of parametric instability is due to subharmonic resonance, which leads to unstable modes of double the period of the time-periodic flow. If this instability leads to a supercritical bifurcation, then, as R increases further, in general a stable *quasi-periodic flow* (that is, one with *two or more* fundamental periods) will develop. This occurs, for example, when wavy Taylor vortices become unstable. (Sometimes, with subharmonic resonance, a stable flow of double the period of the original periodic flow may develop. This may occur for Rayleigh–Bénard convection in tall cells.) As R increases further, more instabilities and bifurcations may occur in which a stable flow with three fundamental frequencies arises. Alternatively, or as R increases yet further, *chaotic* flow may develop; for chaotic flow the Fourier spectrum shows not sharp peaks at the fundamental frequencies but a broad band of frequencies, and the correlation between the velocity at a given point at one time and another diminishes rapidly as the time difference increases. (For the Fourier spectrum we take the time series of a given flow quantity, for example a velocity component at a given point, for a long time and take its Fourier transform to find the strength of the component of each frequency.)

The stable flow following equilibration after the first bifurcation of a basic flow as the Reynolds number increases is often called the *secondary flow* in physical descriptions; and its instability at the second bifurcation as the Reynolds number increases is often called *secondary instability*. Then the basic flow before the first bifurcation would be called the *primary flow* and the equilibrated flow after the primary instability the *secondary flow*. For example, if the first instability leads to equilibration as a standing or travelling wave, and that wave becomes unstable to side bands or some other resonant wave interaction, then there is a secondary instability. Often the primary instability of a two-dimensional flow leads to equilibration as a stable steady two-dimensional wave (more often quasi-steady in practice), but three-dimensionality arises at the secondary instability, where the two-dimensional symmetry is broken. The occurrence of a quasi-periodic flow following the development of a periodic flow as a secondary flow at a Hopf bifurcation of the basic flow is another common form of secondary instability, again exemplified by the occurrence of wavy Taylor vortices.

Successive bifurcations usually follow one another after smaller and smaller increments in the Reynolds number, so as to become indistinguishable in practice. Nonlinear self-interactions of unstable modes generate harmonics and nonlinear mutual interactions of modes generate subharmonics as well. So as the temporal complexity of the flow increases in this sequence of bifurcations when R increases, the spatial complexity usually increases as well. Motion occurs on smaller length scales and on a broader range of length scales. It seems plausible that, for any given class of dynamically similar flows characterized by a Reynolds number, when R is sufficiently large there is a unique statistically stationary turbulent flow independent of the initial conditions; this would be so if there were a unique attractor for sufficiently large values of R.

If a flow has some symmetries, then usually these are at first broken, one by one, at bifurcations as R increases, and finally 'mended' on the onset of turbulence such that the *mean* turbulent flow has all the symmetries (King & Stewart, 1991); an example of this is the breaking of the symmetries of rotation and translation of Couette flow between long coaxial cylinders. Indeed, Couette flow is an excellent prototype of flows with sequences of bifurcations towards transition as the Reynolds number increases; Coles (1965) and Fenstermacher *et al.* (1979) have made classic experimental studies of it.

This description of transition from the point of view of the theory of dynamical systems gives valuable insight, but it is at the risk of over-simplification, and a complementary description of transition from a physical point of view is vital. However, the physical mechanisms vary a lot from class to class of basic flows. Also the mechanisms of transition for some basic flows, notably parallel and nearly parallel flows, are subtle, and the practical realization of a flow is often much more complicated than its model idealized by theoreticians. So transition has to be described physically for each different class of basic flows, and there is only a little more to be written generally.

Transition, especially for parallel or nearly parallel flows, by means other than exponential growth of normal modes is called 'bypass transition', a concept recognized and a phrase coined by Morkovin (1969). Bypass transition may be due to subcritical instability when there is no stable flow contiguous to the basic flow in phase space at the prevailing value of the Reynolds number. It may be due to instability to a finite-amplitude perturbation when the basic flow is stable to all infinitesimal perturbations; for example, when the strong transient growth of a small perturbation of a boundary layer leads to a streaky disturbance elongated downstream which triggers rapid nonlinear amplification.

9.2 Routes to Chaos and Turbulence

We have just related how the transition to turbulence comes first as a sequence of bifurcations. In the final stages of hydrodynamic instability there is onset of chaotic flow with low degrees of freedom, for example, thermal convection in a box heated from below. It is possible to classify roughly various types of turbulence. 'Phase turbulence' with a high degree of freedom involves a range of spatial scales as well as being chaotic in time, for example, thermal convection in a thin layer heated strongly from below. Classical 'shear-flow turbulence' is chaotic in time and space, involving a broad spectrum of wavenumbers yet with a coherent spatial structure, for example, turbulence in a channel or a pipe at fairly large values of the Reynolds number. Finally we meet turbulence at very high values of the Reynolds number, which is chaotic in time and space, with a very broad wavenumber spectrum such that on small scales the flow is isotropic with an inertial range and a viscous cut-off, for example, turbulence downstream of a grid in a large wind tunnel with fast flow. However, hydrodynamic instability strictly concerns only the development up to the onset of phase turbulence, so we will not consider the later states of transition further.

We have just discussed some routes to turbulence as the Reynolds number R increases. Some order may be detected in the jumble of bifurcations. It seems that there are only a few main routes to turbulence, or at least to chaotic flow. So a brief summary with a few more details may help:

(1) *Subcritical instability.* On this route a stable flow, steady, periodic or quasi-periodic, becomes unstable as R increases slowly through a critical value, and the flow then 'jumps' rapidly to a turbulent one which is not a continuous extension of the stable flow for smaller values of R. This occurs, for example, at a pitchfork bifurcation with a *negative* Landau constant; then there is nowhere nearby in phase space for the solution to go to, so it must change substantially and may develop rapidly into turbulence (or it may jump to another stable steady or time-periodic flow, with hysteresis if R increases and then decreases). Such a development of turbulence has been variously called 'abrupt', 'fast' and 'savage'. We find for Poiseuille flow in a pipe, plane Poiseuille flow, plane Couette flow and other channel flows that there is such an abrupt onset of turbulence. For such flows this is often called bypass transition.

(2) *Ruelle–Takens–Newhouse route.* This route was first mapped by Ruelle & Takens (1971) and later revised by Newhouse *et al.* (1978) by use of the theory of dynamical systems. Along this route, there is a succession of bifurcations as R increases in which a steady flow may directly, or via other steady flows, become time-periodic, then quasi-periodic with two and then perhaps three or even four frequencies, until a chaotic flow, which may be regarded as ordered, if not fully developed turbulence, occurs. It seems that quasi-periodic solutions with five or more fundamental frequencies are not stable, and therefore such flows are not observed. As R increases, the fundamental frequencies, say ω_1, ω_2, may vary and so become rationally related, so that ω_2/ω_1 is the ratio of two integers; then the frequencies become 'locked', that is, the flow reverts to a periodic one and stays periodic

as R increases further until it becomes unstable and a new quasi-periodic flow arises. Thus ω_2/ω_1 may increase with R by 'fits and starts' until chaos finally ensues.

This is a very common route for flows with the Landau constant positive, for example, Couette flow between rotating cylinders and Rayleigh–Bénard convection. It is a 'slow' transition by spectral evolution as one bifurcation follows after another until eventually turbulence, or at least chaos, ensues. Sometimes, as the Taylor number increases, chaos arises and then is replaced by a quasi-periodic flow before turbulence eventually ensues.

(3) *Period doubling.* Sometimes turbulence, or chaos rather, ensues after a sequence of *period doubling* bifurcations as R increases. Here a sequence of time-periodic flows occur at bifurcations, the period of one flow being twice the period of the previous one. This has been observed in Rayleigh–Bénard convection in a tall box by Libchaber & Maurer (1978), having been predicted quite generally by Feigenbaum (1980) on theoretical grounds.

(4) *Intermittent transition.* On this route, first mapped mathematically by Pomeau & Manneville (1980), a periodic flow becomes unstable as R increases through a critical value, the stable periodic flow first becoming unstable to disturbances of smaller and smaller finite amplitude. At the critical value R_c of R the stable periodic flow coalesces with an unstable periodic flow (not observable, of course). For small positive $R - R_c$ the same periodic solution appears to persist most of the time (although it no longer corresponds to an exact solution of the governing equations), but is occasionally and rarely interrupted by 'bursts' which are not small. As $R - R_c$ increases further, the bursts occur more frequently but do not change much in magnitude. The average time between bursts tends to infinity like $(R - R_c)^{-1/2}$ as $R \to R_c +$. This theoretical idea also seems to represent turbulent spots and relaminarization.

Exercises

9.1 *What determines the critical value of the dimensionless parameter for onset of instability?* Consider the following physical argument that might be used before calculating stability characteristics. 'The dimensionless parameter, such as a Reynolds or Rayleigh number, specifying a class of dynamically similar flows usually represents a characteristic ratio of destabilizing and stabilizing forces. So the critical value of the parameter for the onset of instability of the basic flow which it specifies might be expected to be of the order of magnitude of 1.' Of course there is no reason to expect the critical value be exactly 1, because the definition of a dimensionless parameter is usually a little arbitrary. Yet, in the problems discussed in this book, the critical values vary from zero to infinity, and many are of the order of magnitude of a thousand. Why do such large critical values occur so often?

[This question is intended to provoke thought rather than evoke *the* correct answer. It might be helpful to consider the changes of the critical values of the Rayleigh number on replacing the horizontal planes $z_* = 0$, d by $z_* = 0$, πd in Chapter 6 on Rayleigh–Bénard convection.]

10

Case Studies in Transition to Turbulence

For I have given you an example, that you should do as I have done
to you.

John xii 15

10.1 Synthesis
10.1.1 Introduction

The plan of this text has to been to describe the important general concepts
and methods of hydrodynamic stability in the opening chapters, and then to
apply them to selected flows in the later chapters. The flows have been selected
partly for their mathematical simplicity, partly for their historical importance
(and these two reasons are connected), and partly for their physical value.
Many of the resultant problems are very idealized; yet all of the problems
are much more widely applicable than their precise form might at first sight
suggest. The theory of Rayleigh–Bénard convection, for example, may be used
to interpret not just instability of an infinite thin horizontal layer of fluid heated
below, but many convective instabilities of flows which *locally* resemble a
thin layer of fluid heated from below. The theory of Taylor vortices may be
used to interpret instabilities of flows with curved streamlines such that there
is a local centrifugal force. The theory of Görtler vortices can be applied to
interpret the local instabilities of flows whose streamlines are convex, so that
this mechanism is complementary to the mechanism of Taylor vortices, to be
applied when the streamlines or the wall 'bend the other way'. The theory
of instability of parallel flows, with Rayleigh's inflection-point theorem and
the Orr–Sommerfeld problem, may be used to interpret instabilities of flows
that are nearly parallel, at least locally; indeed, it has already been used to
interpret instabilities of boundary layers, jets and free shear layers. The use of
these idealized problems to interpret instabilities of more complicated flows
is valuable, but is not easy until one has a lot of experience of hydrodynamic
instability.

The classic problems of hydrodynamic stability have been posed and
solved partly because of their simplicity and partly because of their physical
importance. Their simplicity is due to their spatial and temporal symmetries,

215

which lead to the relative tractability of the linear stability problems, involving the solution of ordinary-differential rather than partial-differential eigenvalue problems. Without this reduction to ordinary-differential systems, the problems of hydrodynamic stability could not have been solved before the advent of supercomputers around 1980. The simplicity of these linear problems, in turn, makes tractable weakly nonlinear analyses of instability. However, eventually, in the route of transition to turbulence as the Reynolds number, Rayleigh number, or the like, increases from zero, there arise problems that can only be solved by use of computational fluid dynamics. So the computer has been the 'engine' of many recent advances in the subject. At the same time, an understanding of many mechanical concepts and simpler problems is valuable to interpret the laboratory and computational results.

Without the ability to use simple problems of ordinary-differential equations as models, we have perforce to rely on laboratory and computational experiments together with the qualitative theory of dynamical systems and miscellaneous physical arguments. These arguments may use simpler problems of hydrodynamic stability, for which ordinary-differential models are available, as local approximations.

Having written these general remarks, we follow with accounts of the instabilities of two important flows for which the problems of linear stability may easily be posed but may be solved only by the use of computational fluid dynamics. The accounts will use whatever methods of investigation are available to further our understanding of the instabilities and transition to turbulence.

10.1.2 Instability of flow past a flat plate at zero incidence

The use of physical arguments and common sense is invaluable, but it may occasionally lead to confusing conclusions when deep issues are involved. To illustrate this, let us revisit the instability of Blasius's boundary layer on a flat plate, briefly putting the boundary layer in the context of the flow as a whole (see Figure 10.1). Next look at Van Dyke (1982, Figs. 29–30) and Homsy *et al.* (CD2000, *Video Library*, 'Flow Past a Flat Plate').

Consider then the idealized thought-experiment of the instability of the steady two-dimensional flow of an infinite uniform stream of an incompressible fluid of kinematic viscosity v with velocity V past a long thin flat plate of chord length L at zero angle of incidence. By dynamical similarity, such flows are characterized by a single dimensionless parameter, say the global Reynolds number $Re = VL/v$. To be sure, there may be hidden variables in a real experiment due to imperfections such as the non-zero thickness of the plate, the shape of its leading and trailing edges, its curvature, the slight

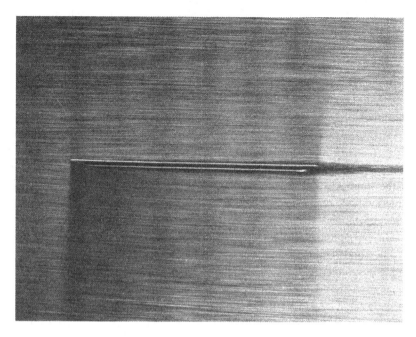

Figure 10.1   Side view of the flow of a uniform stream past a flat plate at zero angle of incidence when $Re = 10\,000$: boundary layer and wake. (After Van Dyke, 1982, Fig. 29; reproduced by permission of ONERA.)

inclination of the plate to the oncoming stream, the inlet and outlet conditions of the finite wind tunnel, the compressibility of the air, the turbulence level in the incident stream, or even the breakdown of the Navier–Stokes equations at molecular length scales. But it is clear from the discussions in this book that in the idealized experiment there will be a sequence of bifurcations as Re increases from zero to infinity, beginning with a symmetric two-dimensional flow as a globally stable attractor, leading to the primary instability (as in Chapter 2), and ending with chaotic vorticity and strongly three-dimensional turbulence. The evidence suggests that the first bifurcation is a supercritical Hopf bifurcation. Perturbations in the wake and the boundary layer on the plate are observed to propagate downstream, so it does seem plausible to consider the wake and boundary-layer instabilities separately in many circumstances, at least as a good approximation. So contrast the above global point of view with the development of the theory of instability of parallel and nearly parallel flows in Chapter 8. There *local* arguments based on boundary-layer theory (which is, of course, only valid if the global Reynolds number Re is large) were used to justify the application of the Orr–Sommerfeld problem without taking into account the length of the plate at all.

In §8.8.5 a local Reynolds number $R = (2Vx_*/\nu)^{1/2}$ based on the thickness $\delta = (2\nu x_*/V)^{1/2}$ of the boundary layer at a chosen station x_* was used, and a critical value of $R_c \approx 520$ was found. How does this determine the critical value of the global Reynolds number Re? It would suggest that instability begins when the local Reynolds number at the trailing edge $x_* = L$ is approximately 520, and so the critical value of Re is of order of magnitude $520^2/2 \approx 135\,000$. For a stream of velocity $V = 1$ m s^{-1} at room temperature and pressure, this gives marginal stability for a plate of length $L \approx 0.14$ m if the fluid is water, or of length $L \approx 2$ m if the fluid is air. So much is described at length in many textbooks on boundary-layer theory (e.g. Batchelor, 1967, §5.8).

Such an argument is admittedly crude, and should only be taken to give order-of-magnitude estimates and qualitative mechanisms. It takes no account of either the complicated structure of the boundary layer at the trailing edge of the plate, or, more importantly, the wake downstream of the plate. The half-width of the wake may be identified with the thickness of the boundary layer at the trailing edge, and so is of the order of magnitude of $(\nu L/2V)^{1/2}$, and the critical value of the Reynolds number of a wake is about 4, the most unstable modes being longish sinuous waves (see §8.7). Then the wake becomes unstable to sinuous perturbations when $V(2\nu L/V)^{1/2}/\nu \approx 4$, that is, $Re \approx 8$, a much lower value than 135 000. This suggests that the instability of the flow around the plate arises from the local wake instability with critical value $Re \approx 8$, not from Tollmien–Schlichting waves on the plate, and that the Tollmien–Schlichting waves arise at much higher values of Re as merely a local instability. Visual observations of the instability of the flow around the plate (Taneda, 1959) give sinuous perturbations aft of the plate at a critical value of Re three or four orders of magnitude less than 135 000, suggesting that the wake instability is the mechanism of the first Hopf bifurcation of the flow as R increases from zero. This is in accordance with the intuition from Chapter 8 that the inviscid mechanism of instability of flows with points of inflection is much stronger than the viscous mechanism of Tollmien–Schlichting instability of flows without. However, the idealized experiment is not easy to approximate closely in the laboratory, and it seems that no definitive critical value of Re has been determined yet. Also, it is clear from the wind-tunnel experiments of Schubauer & Skramstad (1947), Klebanoff *et al.* (1962), Kachanov *et al.* (1975) and many others that the local approximation of the Orr–Sommerfeld problem does represent well the generation of an instability of flow of a low-turbulence stream past a flat plate. For most of these experiments $R \gg Re$, and we use the Orr–Sommerfeld theory not to find whether the flow is unstable so much as *where*, why and how it becomes locally unstable. Near the leading edge the flow is nearly stationary (due to the no-slip condition on the plate), so the *local* Reynolds number is small and the flow is locally well approximated by Stokes flow; it follows that the flow is locally stable (albeit not nearly parallel) for some distance downstream of the leading edge. But as the boundary layer thickens downstream, it becomes nearly parallel and at some station the local Reynolds number reaches its critical value according to the Orr–Sommerfeld problem, and we anticipate that instability grows downstream of that station or thereabouts. Some of this is discussed in §8.5, but the details are complicated and still an active topic of research.

*The apparent paradox whereby the continuum of local modes of instability, in the wake or the boundary layer, say, is related to the discrete *global modes* of a flow as a whole was explained by Stone (1969) and Drazin (1974) for some simple problems. They related the local modes of a slowly varying flow to the global problem by use of the WKBJ approximation. Gent & Leach (1976) applied this idea successfully to the baroclinic instability of flow in a differentially heated rotating annulus by assuming that the flow was nearly parallel. Later, Huerre & Monkewitz (1990) developed it and applied it powerfully to many nearly parallel flows.

10.2 Transition of Flow of a Uniform Stream Past a Bluff Body

The instability of flow of a uniform stream past a bluff body is somewhat similar qualitatively to that past a slender body, but the separation of the flow over a bluff body leads to the development of a 'proper' wake whose cross-section is comparable to the cross-section of the body perpendicular to the stream. In aerodynamics, the boundary layer on a bluff body and its instability are especially important because the position of a point of separation affects the size of the wake, and hence the drag on the body. Indeed, instability is often triggered, say by a trip wire to create perturbations of substantial amplitudes, in order to make the boundary layer turbulent further upstream and thereby reduce the adverse pressure gradient in the boundary layer, delay separation, and thence reduce the drag. Let us look at two examples of flows of a uniform steady stream of an incompressible viscous fluid past bluff bodies.

10.2.1 Flow past a circular cylinder

First let us study the instability of the flow of a uniform stream of an incompressible viscous fluid about a circular cylinder. You cannot start the study

Figure 10.2   The flow of a uniform stream past a circular cylinder at $R = 140$. The flow is from left to right, and the vortex street in the wake of the cylinder is shown. (After Van Dyke, 1982, Fig. 94; reproduced by permission of Professor Sadatoshi Taneda.)

better than by looking at a sequence of pictures of the flow as the Reynolds number increases from zero. You might start by looking again at the pictures in your favorite textbook on fluid dynamics (e.g. Batchelor, 1967, Figs. 4.12.1, 4.12.6, 5.11.4) and looking at Figure 10.2. See also Van Dyke (1982, pp. 4–5, Figs. 6, 24, 40–48, 94–98), Nakayama (1988, pp. iii, x in colour, Figs. 1–9) and Homsy *et al.* (CD2000, *Video Library*, 'Flow Past a Cylinder').

You may note that for very small values of the Reynolds number the flow is symmetric fore and aft of the cylinder as well as 'up and down' in the perpendicular direction. It looks not only steady but also very stable. When the Reynolds number, say $R = VD/\nu$, where V is the velocity of the free stream and D is the diameter of the cylinder, increases to about 1, the fore-and-aft symmetry of the flow has been visibly broken. When R increases to about 10, flow has visibly separated to form a pair of recirculating eddies in the lee of the cylinder. As R increases further, the lines of separation move upstream and the eddies grow longer as a wake, but the flow remains steady and retains its up-and-down symmetry. However, as R increases to R_c, where $R_c \approx 45$, the flow becomes unsteady, the eddies beginning to oscillate. For a stream of velocity $V = 1$ m s^{-1} at room temperature and pressure, this gives marginal stability for a cylinder of diameter $L \approx 0.045$ mm if the fluid is water, or of length $L \approx 0.7$ mm if the fluid is air. The onset of a time-periodic flow indicates a supercritical Hopf bifurcation. As R increases further, vortices are shed alternately from the separated streamlines above and below the cylinder, forming in the wake what von Kármán (1911) called a *vortex street*. The vortex street has the symmetry of a travelling sinuous mode. So much is described at length in many textbooks on boundary-layer theory (e.g. Batchelor, 1967, §§4.12, 5.11) (see Figure 10.3). When R increases to about 200, a secondary instability develops, as two-dimensionality is lost and the vortex street begins to break up. When R becomes about 1000 the flow on, and upstream of, the front of the cylinder still looks laminar, being smooth and fairly steady with a recognizable boundary layer on the cylinder, but the flow in the wake is turbulent, being chaotic, very three-dimensional and strongly diffusive. You can see that the flow on the large scale still retains the character of a vortex street without yet becoming completely disorganized, another example of a coherent structure. As R increases further, the structure of the turbulence becomes finer.

Of course, flow properties of such spatial complexity cannot be easily described by a few ordinary differential equations, so we have to depend upon laboratory and numerical experiments to get quantitative results. However, we can use calculations of the instability of the boundary layer on the cylinder as well as the framework of the qualitative theory of bifurcations to gain some insight into the experimental results. It must be recognized that no laboratory experiment is exactly two-dimensional, and even a numerical experiment cannot be over an unbounded domain of flow, so results of careful experiments may depart a little from the ideal and vary a little among one another according to the aspect ratio of the channel, conditions far from the cylinder, and so forth. Bearing all this in mind, let us again go over the scenario as the Reynolds number increases from zero. More detail will be given this time.

At $R = 0$ there is Stokes flow, which is unique, and therefore has all the symmetries of the configuration – steadiness, two-dimensionality and up-and-down symmetry; it is also time-reversible and so has fore-and-aft symmetry as well. Serrin's theorem shows that the flow is stable (although the theorem strictly applies only to bounded flows) for small values of $R > 0$. For $0 < R < R_c$ there is stable laminar steady two-dimensional flow with up-and-down symmetry. Fore-and-aft symmetry is broken as R increases from zero and a wake is formed, and the wake lengthens as R increases further. At $R = R_c$ the primary flow becomes unstable at a Hopf bifurcation, where steadiness is lost. The up-and-down symmetry *at any given instant* is broken, but the flow at one instant is the mirror image in the centre plane of the flow half a period earlier. Jackson's (1987) numerical calculations of the linear stability characteristics gave $R_c \approx 46$, whereas the laboratory experiments of Provansal *et al.* (1987)

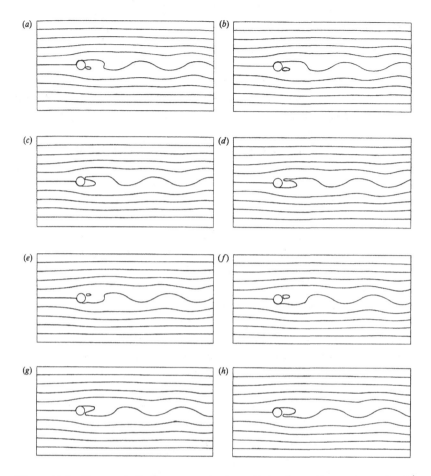

Figure 10.4   Time-periodic flow past a circular cylinder. Streamlines at intervals of $\frac{1}{8}$ of the period when R is a little greater than R_c. (After numerical calculations of Jackson, 1987, Fig. 13.)

for small $R - R_c > 0$ are shown in Figure 10.4. For $R_c < R < R_2$, say, there is a stable laminar two-dimensional time-periodic flow with vortex shedding, the trailing vortices forming a vortex street in the wake of the cylinder. The period of the flow can conveniently be measured by the *Strouhal number*, namely, $St = fD/V$, where f is the frequency of the vortex street. As R increases, St increases slowly from about 0.13. At $R = R_2$ the secondary flow itself becomes unstable, with the onset of three-dimensionality in the vortex street. Experiments show that $R_2 \approx 200$. Williamson (1996) has identified further regimes of flow in higher ranges of Reynolds number.

At $R \approx 260$ another three-dimensional mode of instability in the wake arises. As R increases further, the three-dimensional flow in the wake becomes increasingly disordered, developing a finer-scale structure. At $R \approx 1000$ the detached shear layers in the lee of where the flow separates from the cylinder develop Kelvin–Helmholtz instability, which is principally two-dimensional. The wake of the cylinder remains turbulent, but a coherent structure of a vortex street endures. Williamson also identified the onset of local instability in the boundary layers on the cylinder when $R \approx 200\,000$. It can be seen that even for such a seemingly simple flow there is a very complicated scenario of instabilities and transition as the Reynolds number increases. The results of the numerical and laboratory experiments do not fit any theory as closely as a mathematician might wish, but the route to turbulence does seem to fit the model of Ruelle–Takens–Newhouse.

Again, perturbations propagate downstream mostly, so that it makes sense to talk of the instability of the boundary layer on the cylinder separately from the instability of the wake. Indeed, the whole problem of instabilities of flows about a bluff body is so intractable by methods other than numerical and laboratory experiments, that it is very desirable to decouple the instabilities of the wake and the boundary layers in order to gain some insight into transition. The boundary layer on a curved body such as a cylinder is more complicated than Blasius's boundary layer on a flat plate because the pressure gradient varies downstream, but it is found both observationally and theoretically that at high values of the Reynolds number, separation on the cylinder occurs at angles $\pm\theta$ from the fore stagnation point, where $\theta \approx 80°$. Now, by the theory of irrotational flow about the cylinder, the flow external to the boundary layer on the cylinder is $2V \sin\theta$ at angle θ from the fore stagnation point. This gives crudely a boundary layer of thickness of order of magnitude $\delta = (\nu D\theta / V)^{1/2}$. Therefore the local Reynolds number reaches the critical value 520 at the two points of separation when $R \approx 135\,000$, again, in rough agreement with observations.

10.2.2 Flow past a sphere

Next let us study the instabilities of the flow of a uniform stream past a sphere. You might start by looking again at the pictures in your favorite textbook on fluid dynamics (e.g. Batchelor, 1967, Figs. 4.12.8, 5.11.7). Next look at the pictures of Van Dyke (1982, Figs. 7, 8, 26–28, 49–58) and Homsy *et al.* (CD2000, *Video Library*, 'Flow Past a Sphere' and *Boundary Layers*, 'Instability, Transition, and Turbulence' with sub-subheadings 'Flow over spheres: effect of Re' and 'Tripping the boundary layer'). The scenario of the development of instabilities as the Reynolds number increases from zero is reminiscent of the scenario for flow past a cylinder. But, when interpreting the pictures, remember that the configuration of the flow past a sphere is axisymmetric, not two-dimensional. So there is a steady secondary flow and, not a vortex street, but a periodic shedding of ring vortices as the tertiary flow.

Again, define $R = VL/\nu$, where L is here the diameter of the sphere. When $R = 0$ there is a unique stable steady Stokes flow, with fore-and-aft symmetry as well as axisymmetry. For $0 < R \ll 1$ there is again a unique stable steady axisymmetric flow, but without fore-and-aft symmetry. Nakamura's (1976) visual observations suggest that there is separation with reversed flow aft of the sphere, which begins when $R \approx 7$, and supercitical stability at the onset of the primary instability at $Re \approx 190$, above which an unsteady secondary flow forms. Kim & Pearlstein (1990) computed the basic steady flow, linearized the Navier–Stokes equations, and computed the stability characteristics; they found a primary instability, at $Re \approx 175$, the most unstable mode being oscillatory and having azimuthal wavenumber $n = 1$. However, Natarajan & Acrivos (1993) similarly computed the stability characteristics; they found a primary instability, governed by the principle of exchange of stabilities, at $Re \approx 210$, the most unstable mode being steady, concentrated in the toroidal region of separated flow immediately behind the sphere, and having azimuthal wavenumber $n = 1$. The next most unstable mode is oscillatory and becomes unstable when $Re \approx 277$, which Natarajan & Acrivos suggested might be related to the onset of vortex shedding at a secondary instability. Taneda (1978) reported visual observations of the flow at higher values of the Reynolds number.

10.3 Transition of Flows in a Diverging Channel

10.3.1 Introduction

The flow of an incompressible viscous fluid driven steadily along a channel poses a classic problem of fluid mechanics, whose study goes back to the work of Leonardo da Vinci in the fifteenth century. We will review the problem, or rather range of problems, and its applications by use of the results of modern laboratory experiments, numerical simulations and mathematical methods. It is valuable to have a general framework to help us understand the instability of all flows in diverging channels, rather than only detailed laboratory and numerical results for a few specific flows, which tell us only the properties of those specific flows. Mathematical and physical ideas provide that framework both to interpret laboratory and numerical experiments, and to motivate new ones. Sobey (2000, Chap. 10) has described the basic steady flows in a diverging channel by use of interactive boundary-layer theory and of numerical simulation, and also reviewed their bifurcations; their symmetry breaking is an example of what is called the *Coanda effect*.

The theory of dynamical systems suggests that it is valuable to consider the ensemble of dynamically similar flows as the Reynolds number R increases from zero to infinity. Serrin's (1959) theorem shows that there is a unique steady solution when R is sufficiently small, and that it is a global attractor. As the Reynolds number increases, a succession of bifurcations ensues, leading towards chaos. Without identifying chaos with turbulence, we anticipate that a chaotic attractor will represent turbulence when R is sufficiently large.

If the channel has mirror symmetry (and so is invariant under a Z_2 group of transformations) in one or two planes, then first symmetry breaking and later symmetry mending of the flow will be seen in the bifurcations as R increases. Serrin's theorem implies that the flow for a small value of R has all the symmetries of the channel, and it is plausible that the mean turbulent flow for any large enough value of R also has all the symmetries.

The power of these methods can be seen by applying them qualitatively to interpret pictures of laboratory and numerical experiments (see Figures 10.5, 10.6 and 10.7). The occurrence of a supercritical pitchfork bifurcation and symmetry breaking in these experiments is clear. There is evidence of a further pitchfork bifurcation, with symmetry breaking in the spanwise direction and thence the onset of strongly three-dimensional flow, in the laboratory experiments of Sobey & Drazin (1986, Fig. 12).

10.3.2 Asymptotic methods

Asymptotic methods may be used if the channel has either nearly parallel walls or nearly plane walls (that is, walls of small curvature), although neither case covers the practically important sudden expansion in a channel with plane parallel walls. In the former case the flow is locally like plane Poiseuille flow, and the stability characteristics of small perturbations may be found, to leading asymptotic order, by use of the Orr–Sommerfeld problem. This is the classic approximation to the stability of nearly parallel flows such as Blasius's boundary layer, as discussed in §8.5. In the latter case the flow is locally like a Jeffery–Hamel flow (see Example 2.5), and an analogous asymptotic approach to the flow in a diverging or converging channel (Fraenkel, 1962, 1963) and its stability (Sobey & Drazin, 1986) may be made. This application of the assumption that the curvature of the channel walls is small might seem very restrictive. Yet the *qualitative* results of the asymptotic theory are applicable even when the curvature of the walls is not very small, so the theory of Jeffery–Hamel flows is more important than appears at first sight.

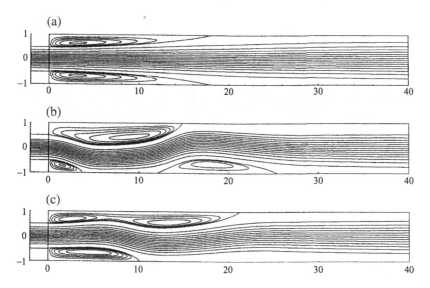

Figure 10.5   Numerical simulation of three steady two-dimensional flows through a sudden expansion from left to right at a largish value of the Reynolds number (so large that one would expect the asymmetric flows as well as the symmetric flow to be unstable). (a) A symmetric flow (unstable). (b) An asymmetric flow. (c) The other asymmetric flow. (After Alleborn *et al.*, 1997, Fig. 8.)

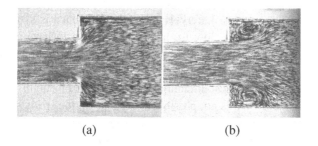

(a)   (b)

Figure 10.6   The flow in a diverging channel. The steady flow of water, from left to right mostly, is visualized by small particles. (a) A low value of R. (b) A five-times-higher value of R. (After Nakayama, 1988, Figs. 114, 115; reproduced by permission of Professor Masanobu Yamamusu.)

Recall from Example 2.5 the classic theory, by Jeffery and Hamel, of the exact solutions of the Navier–Stokes equations which describe the steady two-dimensional radial flow of an incompressible viscous fluid between rigid plane walls. There is an infinity of such flows for each pair of values of the governing dimensionless parameters, namely, the semi-angle α between the

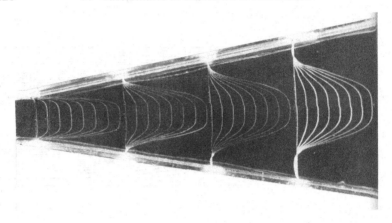

Figure 10.7   The flow through a diverging channel with plane walls with semi-angle
$\alpha = 10°$ at $R = 300$. The steady flow of water, from left to right mostly, is visualized
by a line of hydrogen bubbles produced at eight one-second intervals at each of four
stations. (After Nakayama, 1988, Fig. 105; reproduced by permission of Professor
Yasuki Nakayama.)

two inclined walls and the Reynolds number

$$R = Q/2\nu,$$

where Q is the volume flux per unit length of the line source at the intersection
of the walls. The primary Jeffery–Hamel solution, which is symmetric and
becomes plane Poiseuille flow for $\alpha = 0$, was shown to have a subcritical
pitchfork bifurcation with symmetry breaking as R increases, for fixed $\alpha \neq 0$,
through a critical value, called $R_2(\alpha)$.

There are severe technical difficulties in the problem of stability of Jeffery–
Hamel flows, because it is not possible to separate all the spatial variables as
well as time and thereby reduce the linear stability problem to an ordinary-
differential system. Eagles (1966) used the approximation (mentioned at the
beginning of this section) of nearly parallel flow to investigate the stability
of a Jeffery–Hamel flow to temporal normal modes in the limit as $\alpha \to 0$.
Assuming a locally parallel flow at any given station, he found, by solving
the Orr–Sommerfeld problem, that increase of α from zero (when the Jeffery–
Hamel flow is simply plane Poiseuille flow) to a small positive value is strongly
destabilizing. *This* destabilization is due entirely to the change of the basic
velocity profile, because Eagles neglected the divergence of the streamlines.
However, Banks *et al.* (1988, §5) have questioned the conceptual basis of using
standard normal modes in a sector without excluding the regions corresponding to $r = 0, \infty$.

Dean (1934) found that there is also a separable linear ordinary-differential eigenvalue problem for *steady two-dimensional spatial* modes, on taking a mode with perturbation streamfunction of the form

$$\psi'(r, \theta) = \text{Re}\big[r^\lambda \phi(\theta)\big]. \tag{10.1}$$

Here r, θ are plane polar coordinates, and λ, ϕ are to be found as an eigensolution. It follows that the mode grows or decays in space like $\exp[\text{Re}(\lambda) \log r]$ while it oscillates sinusoidally like $\exp[\text{i Im}(\lambda) \log r]$. The eigenvalue problem may now be shown, by taking $\psi = \Psi + \psi'$ and linearizing the vorticity equation (2.4) for small perturbations ψ', to be

$$\phi^{iv} + \big[\lambda^2 + (\lambda - 2)^2\big]\phi'' + \lambda^2(\lambda - 2)^2\phi$$
$$= R\big[(\lambda - 2)\Psi'\big(\phi'' + \lambda^2\phi\big) - 2\Psi''\phi' - \lambda\Psi'''\phi\big], \tag{10.2}$$

$$\phi(\theta) = \phi'(\theta) = 0 \quad \text{at } \theta = \pm\alpha, \tag{10.3}$$

where Ψ is the streamfunction of the basic flow and a prime here denotes differentiation with respect to θ. We call this the *Dean problem*. There are two countably infinite families of eigenvalues λ with corresponding eigenfunctions ϕ. The real part of λ gives the algebraic rate of growth or decay of the mode as a function of r. A condition of spatial stability seems (Banks *et al.*, 1988) to be that no disturbance grows down- or upstream, so that $\text{Re}(\lambda) \leq 0$ for all modes of one family and $\text{Re}(\lambda) \geq 2$ for all modes of the other family for given Ψ, R, α. (Although stability has been defined in terms of temporal growth of small perturbations, a flow is sometimes said to be spatially stable if all steady small perturbations decay spatially.) This gives a criterion for instability to *steady* two-dimensional perturbations, but, of course, tells nothing about temporally oscillating perturbations, that is, about normal modes not governed by the principle of exchange of stabilities. It also tells nothing about three-dimensional perturbations. Banks *et al.* (1988) found in this way that divergence of the flow is a strongly destabilizing influence but convergence strongly stabilizing. Compare the two photographs in Figure 10.8.

However, McAlpine & Drazin (1998) satisfied the linearized vorticity equation for unsteady two-dimensional flow asymptotically for large r by taking modes of the form

$$\psi'(r, \theta, t) = \text{Re}\big\{\exp\big[ik\big(\alpha^{-1} \log r - ct/\alpha^2 r^2\big)\big] f(y)\big\}, \tag{10.4}$$

where $y = \theta/\alpha$. The motivation for their assumption, that the exponent is inversely proportional to the square of r as well as proportional to t, is the need to balance the terms $\partial\zeta'/\partial t$ and $\Delta\zeta'$ in the vorticity perturbation equation

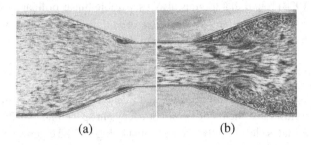

(a)   (b)

Figure 10.8   (a) The flow in a sudden contraction. (b) The flow in a sudden expansion. (After Nakayama, 1988, Figs. 112, 113; reproduced by permission of Professor Masanobu Yamamusu.)

(where ζ' is the perturbation vorticity). Also the scaling of c by the factor α^2 is convenient in order to derive later the Orr–Sommerfeld equation smoothly in the limit as $\alpha \to 0$. This separated solution does not satisfy the linearized vorticity equation exactly, but satisfies it asymptotically as $r \to \infty$ for fixed t, y. It follows that

$$f^{\text{iv}} - \left[k^2 + (k + 2i\alpha)^2\right]f'' + k^2(k + 2i\alpha)^2 f$$
$$= iR\big\{[(k + 2i\alpha)U - kc]\big(f'' - k^2 f\big) - kU''f$$
$$+ 2i\alpha U' f' + 4i\alpha kc(k + i\alpha)f\big\}, \tag{10.5}$$

where now a prime denotes differentiation with respect to y and U is defined by $U(y) = \alpha d\Psi(\theta)/d\theta$. Also the boundary conditions give

$$f(y) = f'(y) = 0 \quad \text{at } y = \pm 1. \tag{10.6}$$

Note that when $c = 0$ and we identify $\lambda = ik/\alpha$, equation (10.5) is equivalent to the Dean equation (10.2), and when $\alpha = 0$ it reduces to the Orr–Sommerfeld equation,

$$f^{\text{iv}} - 2k^2 f'' + k^4 f = ikR\big[(U - c)\big(f'' - k^2 f\big) - U''f\big]. \tag{10.7}$$

Further, note that equation (10.4) gives a classic Orr–Sommerfeld mode of the form

$$\psi'(r, \theta, t) = \text{Re}\big[e^{ik(x-ct)}f(y)\big]$$

in the limit as $\alpha \to 0$ for fixed dimensionless downstream distance $x = (r - r_0)/\alpha r_0$ and local half-width αr_0 of the channel at station $r = r_0$. Thus this generalizes the Orr–Sommerfeld and Dean stability problems for channels with nearly plane walls.

McAlpine & Drazin (1998) solved the eigenvalue problem (10.5), (10.6) numerically in a few cases, and confirmed the earlier conclusions of the strong stabilizing influence of convergence and the strong destabilizing influence of divergence in a channel. In this problem the Tollmien–Schlichting waves of the Orr–Sommerfeld problem change continuously into the Dean modes as the wavenumber of the perturbation decreases, and are thereby linked with the pitchfork bifurcation of Jeffery–Hamel flows. Also the results elucidate the great change in the *mechanism* of instability due to even slight nonparallelism of the channel walls.

10.3.3 Some paradoxes

By applying the theory of Jeffery–Hamel flows to a channel with nearly plane walls, Sobey & Drazin (1986) suggested that the first bifurcation with symmetry breaking occurs when $R = R_c$, where

$$R_c \approx R_2 \left(\tfrac{1}{2} \max_x [\alpha(x) + \beta(x)] \right).$$

α, β are here defined as the angles that the tangent planes to the channel walls at station x make with the (x, z)-plane, the x-axis is directed down the channel, and the z-axis spanwise. (Recall that R_2 is the value of R where the symmetric flow first becomes unstable, at the pitchfork bifurcation of Jeffery–Hamel flows.) This criterion for stability seems to have some support from numerical experiments, though no experiment has been designed to test the criterion when it should be valid (for channels with walls of very small curvature). The theory of spatial stability (Banks *et al.*, 1988) suggests that, when $R > R_c$, Saint-Venant's principle breaks down, whereby small steady perturbations of the flow at either the inlet or the outlet may have substantial effects on the flow in the whole channel. This conclusion is supported by the numerical results of Dennis *et al.* (1997). It suggests that experimentalists should specify carefully their inlet and outlet conditions, because these conditions may strongly influence their observations, and so need to be known to repeat the experiments.

Yet for $R > R_c$ there is no symmetric or asymmetric Jeffery–Hamel flow to represent the observed steady channel flows (see Figure 10.5(b)). Further, Jeffery–Hamel flows exhibit their first instability and symmetry breaking at a *sub*critical pitchfork bifurcation, whereas laboratory and numerical experiments exhibit their first instability and symmetry breaking at a *super*critical pitchfork bifurcation, although these experiments have not been made for small semi-angles α between the walls of the channel. Hamadiche *et al.* (1994) made a specific numerical investigation of this instability for Jeffery–Hamel flow in a channel which is a sector bounded by arcs of small and large radii, and found instability at a supercritical Hopf bifurcation for a range of values of α. These remarks suggest that Jeffery–Hamel flows are irrelevant to flow in a channel if $R > R_c$.

Again, plane Poiseuille flow is well known, by solving the Orr–Sommerfeld problem, to have a subcritical *Hopf* bifurcation at a Reynolds number $R_c \approx 3848$. (Note that the Reynolds number defined here is two-thirds of the Reynolds number defined in §8.7, and $\frac{2}{3} \times 5772 = 3848$.) However, the corresponding approximate Jeffery–Hamel flow has been shown, by solving the Dean problem, to have a subcritical *pitchfork* bifurcation when $R = 3848$ if the semi-angle α is 0.07 degrees. This is a very small angle, not far above the tolerance to be expected in the manufacture of a channel designed to have parallel walls. McAlpine & Drazin (1998) have gone some way to synthesize these two distinct modes of instability associated with the Orr–Sommerfeld and Dean problems, but do not explain why a supercritical pitchfork bifurcation occurs in a diverging channel. One might add that it is plausible that if the walls of a channel are nearly parallel, that is, if $\max_x[\alpha(x) + \beta(x)] \ll 1$, then there is a Hopf rather than a pitchfork bifurcation, but there is no experiment and no theory at present to describe this qualitative change of behaviour as $\max_x[\alpha(x) + \beta(x)]$ decreases to zero.

10.3.4 Nonlinear waves

Tutty (1996) discovered some remarkable waves in a numerical study of some steady two-dimensional flows in a channel, composed of a main section with plane walls (as in a Jeffery–Hamel flow), but with more complicated inlets and outlets. At a few pairs of values of the semi-angle α between the plane walls and of the Reynolds number R, he found flows which seem to be approximately periodic in $\log r$ for a long distance r downstream (Tutty, 1996, Figs. 7–10). Such a steady wave has a streamfunction which depends on r as well as θ, coexisting with some Jeffery–Hamel flows, whose streamfunctions depend only on θ. It would seem from Tutty's numerical experiments that these waves are stable, at least to two-dimensional steady perturbations at the appropriate values of α, R.

Kerswell *et al.* (2002) have elucidated these steady strongly nonlinear two-dimensional wave perturbations of Jeffery–Hamel flows by numerical calculations. They have unfolded their complicated bifurcation structure by a path-following method. Note that a short train of similar waves is observed (see Figure 10.5(b)) to develop, as the Reynolds number increases above its critical value for the first symmetry-breaking bifurcation, downstream of the expansion sections of channels in other experiments we have noted. So Tutty waves mark an interesting advance from Jeffery–Hamel flows and their instabilities towards chaos and transition to turbulence in flow in a diverging channel.

10.3.5 Conclusions

In summary, it may be said that we have described substantial progress in the understanding of the early stages of transition of channel flows made during the last two decades, but that understanding of the later stages is still poor. This progress makes it practical to interpret qualitatively the chief properties of flow in a wide variety of channels, provided that the Reynolds number is not large. The ideas described illustrate more generally the challenge of understanding the instability of nonparallel flows. Indeed, these problems of bifurcation and instability of channel flows comprise an instructive case study of hydrodynamic stability.

Exercises

10.1 *The instability of a wake and formation of a vortex street.* You are given that the steady two-dimensional flow of a uniform stream of a viscous incompressible fluid about a bluff body forms a wake, which becomes unstable at a Hopf bifurcation as the Reynolds number increases above a critical value. The bifurcated flow is seen at moderate values of the Reynolds number as a vortex street shed from the rear of the body as a pair of staggered rows of vortices (with equal and opposite circulation when the body is symmetric about a plane parallel to the uniform stream) which propagates steadily downstream.

Von Kármán (1911) modelled a vortex street in the wake of a symmetric body as irrotational flow of an *inviscid* fluid about *line* vortices of circulation κ at points $(ma + Vt, \frac{1}{2}b)$ and $-\kappa$ at $((n + \frac{1}{2})a + Vt, -\frac{1}{2}b)$ in the (x, y)-plane, for $m, n = 0, \pm1, \pm2, \ldots$ for given constant distances $a, b > 0$ and some steady velocity V.

Using the theory of plane irrotational flow, show that the velocity of von Kármán's vortex street is given by

$$V = (\pi\kappa/a)\tanh(\pi b/a).$$

Deduce that the unsteady flow has period $a^2/\pi\kappa \tanh(\pi b/a)$ in time as well as a in the x-direction. Sketch the streamlines at $t = 0$. [Lamb (1932, §156).]

10.2 *Instability of a vortex street.* Taking the vortex sheet in Exercise 10.1 above as the basic flow, let the vortices be displaced to $(ma + Vt + x_m, \frac{1}{2}b + y_m)$, $((n + \frac{1}{2})a + Vt + x'_n, -\frac{1}{2}b + y'_n)$. Show (by Floquet theory of instability of periodic solutions, if you know it) that it is plausible that each linear perturbation is a superposition of modes of the form

$$x_m = \alpha e^{im\phi}, \qquad y_m = \beta e^{im\phi}, \qquad x'_n = \alpha' e^{in\phi}, \qquad y'_n = \beta' e^{in\phi},$$

where $0 \le \phi < 2\pi$ without loss of generality, and $\alpha, \alpha', \beta, \beta'$ are some functions of t. [If $\phi = 0$, the mode has the same x-wavelength a as the basic flow; if $\phi = \pi$, the mode is a subharmonic, but the phase change ϕ of a mode over the wavelength a may be any angle.]

Assuming two-dimensional irrotational flow about the line vortices, linearizing the infinite system of ordinary differential equations which governs the coordinates of each vortex in the two rows of vortices with respect to small displacements, show that

$$\frac{a^2}{\kappa}\frac{d\alpha}{dt} = -A\beta - B\alpha' - C\beta', \qquad \frac{a^2}{\kappa}\frac{d\beta}{dt} = -A\alpha - C\alpha' + B\beta',$$

$$\frac{a^2}{\kappa}\frac{d\alpha'}{dt} = A\beta' - B\alpha + C\beta, \qquad \frac{a^2}{\kappa}\frac{d\beta'}{dt} = A\alpha' + C\alpha + B\beta,$$

where

$$A = \tfrac{1}{2}\phi(2\pi - \phi) - \pi^2 \mathrm{sech}^2(k\pi),$$

$$B = i\big[\pi\phi \sinh k(\pi - \phi)\mathrm{sech}(k\pi) + \pi^2 \sinh(k\phi)\mathrm{sech}^2(k\pi)\big],$$

$$C = \pi^2 \cosh(k\phi)\mathrm{sech}^2(k\pi) - \pi\phi \cosh k(\pi - \phi)\mathrm{sech}(k\pi),$$

and $k = b/a$.

Prove that the above mode may be expressed as the sum of two types of independent modes, a symmetric mode with $\alpha' = \alpha$, $\beta' = -\beta$ and antisymmetric mode with $\alpha' = -\alpha'$, $\beta' = \beta$. Taking $\alpha, \beta \propto e^{st}$, show that

$$a^2 s/\kappa = B \pm \left(A^2 - C^2\right)^{1/2},$$

for the antisymmetric mode, and that a similar formula (with B replaced by $-B$) governs the symmetric mode. Deduce that the vortex street is exponentially unstable unless $b = k_c a$ precisely, where $k_c = \pi^{-1}\cosh^{-1}(2^{-1/2}) = 0.2806$. [Lamb (1932, §156).]

10.3 *Evolution of small perturbations of a Jeffery–Hamel flow in space and time.* The vorticity equation for two-dimensional motion of a viscous incompressible fluid is

$$\frac{\partial \zeta}{\partial t} + \frac{1}{r} \frac{\partial(\zeta, \psi)}{\partial(r, \theta)} = \nu \Delta \zeta,$$

where ψ is the streamfunction, the vorticity $\zeta = -\Delta \psi$ and the Laplacian is $\Delta = \partial^2/\partial r^2 + \partial/r\partial r + \partial^2/r^2\partial\theta^2$ in terms of plane polar coordinates (r, θ). Taking small two-dimensional perturbations of a radial flow, put

$$\psi = \tfrac{1}{2} Q(\Psi + \psi'), \tag{E10.1}$$

where $\Psi(\theta)$ is a solution of the *Jeffery–Hamel problem* (see Example 2.5), linearize the vorticity equation and the boundary conditions at fixed rigid walls $\theta = \pm\alpha$, and show that

$$\Delta\zeta' = R\frac{\partial\zeta'}{\partial t} + \frac{R}{r}\frac{d\Psi}{d\theta}\frac{\partial\zeta'}{\partial r} + \frac{R}{r^3}\frac{d^3\Psi}{d\theta^3}\frac{\partial\psi'}{\partial r} + \frac{2R}{r^4}\frac{d^2\Psi}{d\theta^2}\frac{\partial\psi'}{\partial\theta}, \tag{E10.2}$$

$$\psi'(r, \theta, t) = \frac{\partial\psi'}{\partial\theta}(r, \theta, t) = 0 \quad \text{at } \theta = \pm\alpha, \tag{E10.3}$$

where the vorticity perturbation is $\zeta' = -\Delta\psi'$, the Reynolds number is defined as $R = Q/2\nu$ and dimensionless variables are used (see Exercise 2.13).

Taking modes of the form

$$\psi'(r, \theta, t) = \text{Re}\{\exp[ik(\alpha^{-1}\log r - ct/\alpha^2 r^2)]f(y)\}, \tag{E10.4}$$

where $y = \theta/\alpha$, k is a real wavenumber and c a complex velocity, show that

$$f^{\text{iv}} - [k^2 + (k + 2i\alpha)^2]f'' + k^2(k + 2i\alpha)^2 f$$
$$= iR\{[(k + 2i\alpha)U - kc](f'' - k^2 f) - kU'' f$$
$$+ 2i\alpha U'f' + 4i\alpha kc(k + i\alpha)f\} \tag{E10.5}$$

in the limit as $r \to \infty$ for fixed t, and that

$$f(y) = f'(y) = 0 \quad \text{at } y = \pm 1, \tag{E10.6}$$

where $U(y) = \alpha d\Psi(\theta)/d\theta$ and a prime now denotes differentiation with respect to y. Verify that the Orr–Sommerfeld equation follows if $\alpha = 0$ and the Dean equation (E2.4) if $c = 0$. [McAlpine & Drazin (1998).]

References

Numbers in square brackets at the end of a reference refer to the pages on which the reference is cited.

Abramowitz, M. & Stegun, I. A. (Editors) (1964) *Handbook of Mathematical Functions*, Appl. Math. Ser. No. 55 (Govt. Printing Office, Washington, D.C.; also Dover, New York, 1965). [39]

Alleborn, N., Nandakumar, K., Raszillier, H. & Durst, F. (1997) Further contributions on the two-dimensional flow in a sudden expansion, *J. Fluid Mech.* **330**, 169–188. [227]

Andereck, C. D., Liu, S. & Swinney, H. (1986) Flow regimes in a circular Couette system with independently rotating cylinders, *J. Fluid Mech.* **164**, 155–183. [132]

Baines, P. G. & Gill, A. E. (1969) On thermohaline convection with linear gradients, *J. Fluid Mech.* **37**, 289–306. [122]

Baines, P. G. & Mitsudera, H. (1994) On the mechanism of shear flow instabilities, *J. Fluid Mech.* **276**, 327–342. [145]

Baines, P. G., Majumdar, S. & Mitsudera, H. (1996) The mechanics of the Tollmien–Schlichting wave, *J. Fluid Mech.* **312**, 107–124. [161]

Banks, W. H. H., Drazin, P. G. & Zaturska, M. B. (1988) On perturbations of Jeffery–Hamel flow, *J. Fluid Mech.* **186**, 559–581. [40, 228, 229, 231]

Barcilon, A. & Drazin, P. G. (2001) Nonlinear waves of vorticity, *Studies Appl. Math.* **106**, 437–479. [191]

Batchelor, G. K. (1967) *An Introduction to Fluid Dynamics* (Cambridge University Press). [16, 39, 46, 158, 182, 198, 220, 221, 225].

Batchelor, G. K. & Gill, A. E. (1962) Analysis of the stability of axisymmetric jets, *J. Fluid Mech.* **14**, 529–551. [179, 182]

Bayly, B. J., Orszag, S. A. & Herbert, T. (1988) Instability mechanisms in shear-flow transition, *Ann. Rev. Fluid Mech.* **20**, 359–391. [176, 177]

Beale, J. T., Kato, T. & Majda, A. (1984) Remarks on the breakdown of smooth solutions for the 3-dimensional Euler equations, *Commun. Math. Phys.* **94**, 61–66. [154]

Bellman, R. & Pennington, R. H. (1954) Effects of surface tension and viscosity on Taylor instability, *Quart. Appl. Math.* **12**, 151–162. [60]

Bénard, H. (1900) Les tourbillons cellulaires dans une nappe liquide, *Revue Gén. Sci. Pure Appl.* **11**, 1261–1271, 1309–1328. [93, 94]

238 References

Benjamin, T. B. (1976) Applications of Leray–Schauder degree theory to problems of hydrodynamic stability, *Math. Proc. Camb. Phil. Soc.* **79**, 373–392. [208]

Benjamin, T. B. & Mullin, T. (1982) Notes on the multiplicity of flows in the Taylor experiment, *J. Fluid Mech.* **121**, 219–230. [130, 209]

Bertolotti, F. P., Herbert, T. & Spalart, P. R. (1992) Linear and nonlinear stability of the Blasius boundary layer, *J. Fluid Mech.* **242**, 441–474. [159]

Betchov, R. & Criminale, W. O. (1967) *Stability of Parallel Flows* (Academic Press, New York). [9]

Boberg, L. & Brosa, U. (1988) Onset of turbulence in a pipe, *Z. Naturforsch.* **439**, 697–726. [207]

Bohr, N. (1909) Determination of the surface-tension of water by the method of jet vibration, *Phil. Trans. Roy. Soc. Lond.* A **209**, 281–317. Also in *Collected Works* **1**, 29–65. [74]

Boussinesq, J. (1903) *Théorie Analytique de la Chaleur* (Gauthier–Villars, Paris). [95]

Bouthier, M. (1973) Stabilité linéaire des écoulements presque parallèles. Partie II. La couche limite de Blasius, *J. Méc.* **12**, 75–95. [159]

Briggs, R. J. (1964) *Electron-Stream Interactions in Plasmas* (M.I.T. Press, Cambridge, MA). [71]

Brown, S. N. & Stewartson, K. (1979) On the secular stability of a regular Rossby neutral mode, *Geophys. Astrophys. Fluid Dyn.* **14**, 1–18. [33]

Burridge, D. M. (1970) The instability of round jets, *Tech. Rep.*, No. 29, Geophys. Fluid Dynamics Inst., Florida State University. [182]

Busse, F. H. & Clever, R. M. (1979) Instabilities of convection rolls in a fluid of moderate Prandtl number, *J. Fluid Mech.* **91**, 319–335. [109]

Busse, F. H. & Whitehead, J. A. (1971) Instabilities of convection rolls in a high Prandtl number fluid, *J. Fluid Mech.* **47**, 305–320. [109]

Butler, K. M. & Farrell, B. F. (1992) Three-dimensional optimal perturbations in viscous shear flow, *Phys. Fluids A* **4**, 1637–1650. [177]

Campbell, L. & Garnett, W. (1882) *Life of James Clerk Maxwell* (Macmillan, London). [1]

Carlson, D. E., Widnall, S. E. & Peeters, M. F. (1982) A flow-visualization study of transition in plane Poiseuille flow, *J. Fluid Mech.* **121**, 487–505. [6]

Case, K. M. (1960) Stability of inviscid plane Couette flow, *Phys. Fluids* **3**, 143–148. [150]

Cesari, L. (1959) *Asymptotic Behaviour and Stability Problems in Ordinary Differential Equations* (Springer, Berlin). [37]

Chandrasekhar, S. (1961) *Hydrodynamic and Hydromagnetic Stability* (Oxford University Press). [9, 60]

Christopherson, D. G. (1940) Note on the vibration of membranes, *Quart. J. Math.* **11**, 63–65. [112]

Coles, D. (1965) Transition in circular Couette flow, *J. Fluid Mech.* **21**, 385–425. [211]

Colson, D. (1954) Wave-cloud formation at Denver, *Weatherwise* **7**, 34–35. [55]

Craik, A. D. D. (1971) Non-linear resonant instability in boundary layers, *J. Fluid Mech.* **50**, 393–413. [177]

Craik, A. D. D. (1985) *Wave Interactions and Fluid Flows* (Cambridge University Press). [81]

Craik, A. D. D. & Criminale, W. O. (1986) Evolution of wavelike disturbances in shear flows: a class of exact solutions of the Navier–Stokes equations, *Proc. Roy. Soc. Lond.* A **406**, 13–26. [42, 186]

Davey, A. (1962) The growth of Taylor vortices in flow between rotating cylinders, *J. Fluid Mech.* **14**, 336–368. [128]

Dean, W. R. (1934) Note on the divergent flow of fluid, *Phil. Mag.* **18**(7), 759–777. [40, 41, 229]

Dennis, S. C. R., Banks, W. H. H., Drazin, P. G. & Zaturska, M. B. (1997) Flow along a diverging channel, *J. Fluid Mech.* **336**, 183–202. [231]

Donnelly, R. J. & Glaberson, W. (1966) Experiments on the capillary instability of a liquid jet, *Proc. Roy. Soc. Lond. A* **290**, 547–556. [65]

Donnelly, R. J. & Simon, N. J. (1960) An empirical torque relation for supercritical flow between rotating cylinders, *J. Fluid Mech.* **7**, 401–418. [131]

Drazin, P. G. (1961) Discontinuous velocity profiles for the Orr–Sommerfeld equation, *J. Fluid Mech.* **10**, 571–583. [202]

Drazin, P. G. (1974) On instability of slowly-varying flow, *Quart. J. Mech. Appl. Math.* **27**, 69–86. [219]

Drazin, P. G. (1975) On the effects of side walls on Bénard convection, *Zeit. angew. Math. Phys.* **26**, 239–243. [113]

Drazin, P. G. (1989) Internal gravity waves and shear instability, pp. 61–73 of *Waves and Stability in Continuous Media*, eds. A. Donato & S. Giambò (Editel, Italy). [42]

Drazin, P. G. (1992) *Nonlinear Systems* (Cambridge University Press). [116]

Drazin, P. G. & Reid, W. H. (1981) *Hydrodynamic Stability* (Cambridge University Press). [9, 42, 44, 57, 58, 59, 61, 66, 67, 91, 100, 101, 114, 121, 122, 128, 129, 130, 131, 133, 137, 166, 168, 172, 177, 194]

Eagles, P. M. (1966) The stability of a family of Jeffery–Hamel solutions for divergent channel flow, *J. Fluid Mech.* **24**, 191–207. [228]

Eliahou, S., Tumin, A. & Wygnanski, I. (1998) Laminar-turbulent transition in Poiseuille pipe flow subjected to periodic perturbation emanating from the wall, *J. Fluid Mech.* **361**, 333–349. [5]

Ellingsen, T. & Palm, E. (1975) Stability of linear flow, *Phys. Fluids* **18**, 487–488. [186]

Feigenbaum, M. J. (1980) The transition to aperiodic behaviour in turbulent systems, *Commun. Math. Phys.* **77**, 65–86. [103, 213]

Fenstermacher, P. R., Swinney, H. L. & Gollub, J. P. (1979) Dynamical instabilities and the transition to chaotic Taylor vortex flow, *J. Fluid Mech.* **94**, 103–123. [129, 211]

Fjørtoft, R. (1950) Application of integral theorems in deriving criteria of instability for laminar flows and for the baroclinic circular vortex, *Geofys. Publ., Oslo* **17**(6), 1–52. [145]

Fraenkel, L. E. (1962) Laminar flow in symmetrical channels with slightly curved walls. I. On the Jeffery–Hamel solutions for flow between plane walls, *Proc. Roy. Soc. Lond. A* **267**, 119–138. [18, 39, 226]

Fraenkel, L. E. (1963) Laminar flow in symmetrical channels with slightly curved walls. II. An asymptotic series for the stream function, *Proc. Roy. Soc. Lond. A* **272**, 406–428. [226]

Gaster, M. (1962) A note on the relation between temporally-increasing and spatially-increasing disturbances in hydrodynamic stability, *J. Fluid Mech.* **14**, 222–224. [200]

Gaster, M. (1974) On the effects of boundary-layer growth on flow stability, *J. Fluid Mech.* **66**, 465–480. [159]

Gent, P. R. & Leach, H. (1976) Baroclinic instability in an eccentric annulus, *J. Fluid Mech.* **77**, 769–788. [219]

Goldstein, R. J. & Graham, D. L. (1969) Stability of a horizontal fluid layer with zero shear boundaries, *Phys. Fluids* **12**, 1133–1137. [98]

Goldstein, S. (1931) On the stability of superposed streams of fluids of different densities, *Proc. Roy. Soc. Lond.* A **132**, 524–548. [194]

Gor'kov, L. P. (1957) Stationary convection in a plane liquid layer near the critical heat transfer point, *Sov. Phys. JETP* **6**, 311–315. Also *Zh. Eksp. Teor. Fiz.* **33**, 402–407 (in Russian). [75]

Görtler, H. (1940) Über eine dreidimensionale Instabilität laminarer Grenzschichten an konkaven Wänden, *Nachr. Ges. Wiss. Göttingen* N.F. **1**, No. 1, 1–26. Translated as 'On the three-dimensional instability of laminar boundary layers on concave walls', *Tech. Memor. Nat. Adv. Comm. Aero., Wash.* No. 1375 (1954). [131, 133]

Gregory, N., Stuart, J. T. & Walker, W. S. (1955) On the stability of three-dimensional boundary layers with application to the flow due to a rotating disc, *Phil. Trans. Roy. Soc. Lond.* A **248**, 155–199. [192]

Gull, S. F. (1975) The X-ray, optical and radio properties of young supernova remnants, *Mon. Not. Roy. Astr. Soc.* **171**, 263–278. [51]

Hall, P. (1983) The linear development of Görtler vortices in growing boundary layers, *J. Fluid Mech.* **130**, 41–58. [134]

Hamadiche, M., Scott, J. & Jeandel, D. (1994) Temporal stability of Jeffery–Hamel flow, *J. Fluid Mech.* **268**, 71–88. [232]

Hamel, G. (1916) Spiralförmige Bewegungen zäher Flüssigkeiten, *Jber. dtsch. Math.-Ver.* **25**, 34–60. [39]

Harris, D. L. & Reid, W. H. (1964) On the stability of flow between rotating cylinders. Part 2. Numerical analysis. *J. Fluid Mech.* **20**, 95–101. [128, 129]

Harrison, W. J. (1908) The influence of viscosity on the oscillations of superposed fluids, *Proc. Lond. Math. Soc.* **6**(2), 396–405. [60]

Haurwitz, B. (1931) Zur Theorie der Wellenbewegungen in Luft und Wasser, *Veröff. Geofys. Inst. Univ. Leipzig* **6**, No. 1. [194]

Heisenberg, W. (1924) Über Stabilität und Turbulenz von Flüssigkeitsströmen, *Ann. Phys., Leipzig* **74**(4), 577–627. Translated as 'On stability and turbulence of fluid flows', *Tech. Memor. Nat. Adv. Comm. Aero., Wash.* No. 1291 (1951). [74, 164, 176]

Helmholtz, H. von (1868) Über discontinuirliche Flüssigkeitsbewegungen, *Monats. Königl. Preuss. Akad. Wiss., Berlin* **23**, 215–228. Also in *Wissenschaftliche Abhandlungen* **1** (1882), 146–157. Translated as 'On discontinuous movements of fluids', *Phil. Mag.* **36**(4) (1868), 337–346. [45, 151]

Helmholtz, H. von (1890) Die Energie der Wogen und des Windes, *Ann. Phys.* **41**, 641–662. Also in *Sitz. Akad. Wiss., Berlin* **23**, 853–872, and *Wissenschaftliche Abhandlungen* **3** (1895), 333–355. Translated as 'The energy of the billows and the wind' by Cleveland Abbe, pp. 112–129 of *The Mechanics of the Earth's Atmosphere* (Smithsonian Institute, Washington, D.C., 1891). [54]

Herbert, T. (1988) Secondary instability of boundary layers, *Ann. Rev. Fluid Mech.* **20**, 487–526. [175]

Herivel, J. W. (1955) The derivation of the equations of motion of an ideal fluid by Hamilton's principle, *Proc. Camb. Phil. Soc.* **51**, 344–349. [155]

Herron, I. H. (1987) The Orr–Sommerfeld equation on infinite intervals, *SIAM Rev.* **29**, 597–620. [157]

Hinch, E. J. (1973) Instabilities due to dissipation, *Eureka* **36**, 17–18. [120]

Hocking, L. M. (1975) Non-linear instability of the asymptotic suction velocity profile, *Quart. J. Mech. Appl. Math.* **28**, 341–353. [197]

Horton, C. W. & Rogers, F. T. (1945) Convection currents in a porous medium, *J. Appl. Phys.* **16**, 367–370. [117]

Howard, L. N. (1961) Note on a paper of John W. Miles, *J. Fluid Mech.* **10**, 509–512. [146, 194]

Howard, L. N. (1964) The number of unstable modes in hydrodynamic stability problems, *J. Méc.* **3**, 433–443. [189]

Huerre, P. & Monkewitz, P. A. (1990) Local and global instabilities in spatially developing flows, *Ann. Rev. Fluid Mech.* **22**, 473–537. [71, 219]

Huerre, P. & Rossi, M. (1998) Hydrodynamic instabilities: open flows, Chapter 2, pp. 81–294, of *Hydrodynamics and Nonlinear Instabilities*, eds. C. Godréche & P. Manneville (Oxford University Press). [8]

Hughes, T. H. & Reid, W. H. (1965) On the stability of the asymptotic suction boundary-layer profile, *J. Fluid Mech.* **23**, 715–735. [197]

Jackson, C. P. (1987) A finite-element study of the onset of vortex shedding in flow past variously shaped bodies, *J. Fluid Mech.* **182**, 23–45. [221, 222, 223, 224]

Jeans, J. H. (1902) Stability of a spherical nebula, *Phil. Trans. Roy. Soc. Lond. A* **199**, 1–53. [44]

Jeffery, G. B. (1915) The two-dimensional steady motion of a viscous fluid, *Phil. Mag.* **29**(6), 455–465. [39]

Joseph, D. D. (1965) On the stability of the Boussinesq equations, *Arch. Rat. Mech. Anal.* **20**, 59–71. [114]

Joseph, D. D. (1968) Eigenvalue bounds for the Orr–Sommerfeld equation, *J. Fluid Mech.* **33**, 617–621. [163]

Joseph, D. D. (1969) Eigenvalue bounds for the Orr–Sommerfeld equation. Part 2, *J. Fluid Mech.* **36**, 721–734. [163]

Joseph, D. D. (1976) *Stability of Fluid Motions* (Springer-Verlag, Berlin). [9, 84]

Kachanov, Y. S. (1994) Physical mechanisms of laminar-boundary-layer transition, *Ann. Rev. Fluid Mech.* **26**, 411–482. [177]

Kachanov, Y. S., Kozlov, V. V. & Levchenko, V. Y. (1975) The development of small-amplitude oscillations in a laminar boundary layer, *Fluid Mech. Sov. Res.* **8** (1979), 152–156. Translated from Russian in *Uch. Zap. TsAGI* **6** (1975), 135–140. [219]

Kármán, T. von (1911) Über den Mechanismus des Widerstandes, den ein bewegter Körper in einer Flüssigkeit erfährt, *Nachr. Ges. Wiss. Göttingen, Math.-phys. Klasse*, 509–517. Also in *Collected Works* **1** (1956), 324–330. [221, 233]

Kelly, R. E. (1967) On the stability of an inviscid shear layer which is periodic in space and time, *J. Fluid Mech.* **27**, 657–689. [177]

Kelvin, W. (1871) Hydrokinetic solutions and observations, *Phil. Mag.* **42**(4), 362–377. Also in *Math. Phys. Papers* **IV** (1910), 69–85. [46, 53, 54, 57, 151]

Kelvin, W. (1880) On a disturbing infinity in Lord Rayleigh's solution for waves in a plane vortex stratum, *Nature* **23**, 45–46. Also in *Math. Phys. Papers* **IV** (1910), 186–187. [147]

Kelvin, W. (1887) Stability of fluid motion – rectilineal motion of viscous fluid between two parallel planes, *Phil. Mag.* **24**(5), 188–196. Also in *Math. Phys. Papers* **IV** (1910), 321–330. [186]

Kerswell, R. R., Tutty, O. R. & Drazin, P. G. (2002) Spatially periodic steady flows of a viscous fluid between two inclined planes (to be published). [232]

Kim, I. & Pearlstein, A. J. (1990) Linear stability of the flow past a sphere, *J. Fluid Mech.* **211**, 73–93. [225]

King, G. P. & Stewart, I. N. (1991) Symmetric and chaos, pp. 257–315 in *Nonlinear Equations in the Applied Sciences*, eds. W. F. Ames & C. F. Rogers (Academic Press, New York). [211]

Klaasen, G. P. & Peltier, W. R. (1985) The onset of turbulence in finite-amplitude Kelvin–Helmholtz billows, *J. Fluid Mech.* **155**, 1–35. [55]

Klebanoff, P. S., Tidstrom, K. D. & Sargent, L. M. (1962) The three-dimensional nature of boundary-layer transition, *J. Fluid Mech.* **12**, 1–34. [173, 177, 219]

Koschmieder, E. L. (1993) *Bénard Cells and Taylor Vortices* (Cambridge University Press). [105, 109, 125, 129, 130]

Krasny, R. (1986) Desingularization of periodic vortex sheet roll-up, *J. Computat. Phys.* **65**, 292–313. [54]

Krishnamurti, R. (1973) Some further studies on the transition to turbulent convection, *J. Fluid Mech.* **60**, 285–303. [103]

Krishnamurti, R. (1975) On cellular cloud patterns. Part 3: Applicability of the mathematical and laboratory models, *J. Atmos. Sci.* **32**, 1373–1383. [110]

Lamb, H. (1932) *Hydrodynamics*, 6th edn. (Cambridge University Press). [43, 44, 190, 233, 234]

Landau, L. D. (1944) On the problem of turbulence, *Dokl. Akad. Nauk SSSR* **44**, 311–314. Also in *Collected Papers*, ed. D. ter Haar (Pergamon, Oxford), 387–391. [20, 74, 176, 208]

Landau, L. D. (1946) On the vibrations of the electronic plasma, *J. Phys. USSR* **10**, 23–34. Also in *Collected Papers*, ed. D. ter Haar (Pergamon, Oxford), 445–460. [71]

Landau, L. D. & Lifshitz, E. M. (1987) *Fluid Mechanics*, 2nd edn. (Addison-Wesley, New York). [9]

Lapwood, E. R. (1948) Convection of a fluid in a porous medium, *Proc. Camb. Phil. Soc.* **44**, 508–521. [117]

Leibovich, S. (1969) Stability of density stratified rotating flows, *AIAA J.* **7**, 177–178. [195]

Leibovich, S. (1979) Waves in parallel or swirling stratified shear flows, *J. Fluid Mech.* **93**, 401–412. [195]

Lewis, D. J. (1950) The instability of liquid surfaces when accelerataed in a direction perpendicular to their planes. II, *Proc. Roy. Soc. Lond. A* **202**, 81–96. [51]

Libchaber, A. & Maurer, J. (1978) Local probe in a Rayleigh–Bénard experiment in liquid helium, *J. Physique Lett.* **39**, L369–372. [103, 213]

Lin, C. C. (1955) *The Theory of Hydrodynamic Stability* (Cambridge University Press). [9, 145, 146]

Lin, S. P. & Reitz, R. D. (1998) Drop and spray formation from a liquid jet, *Ann. Rev. Fluid Mech.* **30**, 85–105. [66]

Lindzen, R. S. & Rambaldi, J. S. (1986) A study of over-reflection in viscous Poiseuille flow, *J. Fluid Mech.* **165**, 355–372. [161]

Lorenz, E. N. (1963) Deterministic nonperiodic flow, *J. Atmos. Sci.* **20**, 130–141. [115]

Lundgren, T. S. (1982) Strained spiral vortex model for turbulent fine structure, *Phys. Fluids* **25**, 2193–2241. [52]

Mack, L. M. (1976) A numerical study of the eigenvalue spectrum of the Blasius boundary layer, *J. Fluid Mech.* **73**, 497–570. [171]

Malkus, W. V. R. (1954) Discrete transitions in turbulent convection, *Proc. Roy. Soc. Lond. A* **225**, 185–195. [208]

Malkus, W. V. R. & Veronis, G. (1958) Finite amplitude cellular convection, *J. Fluid Mech.* **4**, 225–260. [75]

Matkowsky, B. J. (1970) A simple nonlinear dynamic stability problem, *Bull. Amer. Math. Soc.* **76**, 620–625. [76]

Maxwell, J. C. (1876) Capillary action, *Encyclopaedia Britannica*, 9th edn, **5**, 59. Also in *Sci. Papers* **2** (1890), 587. [59]

McAlpine, A. & Drazin, P. G. (1998) On the spatio-temporal development of small perturbations of Jeffery–Hamel flows, *Fluid Dyn. Res.* **22**, 123–138. [229, 231, 232, 235]

Meksyn, D. & Stuart, J. T. (1951) Stability of viscous motion between parallel planes for finite disturbances, *Proc. Roy. Soc. Lond.* A **208**, 517–526. [176]

Meseguer, Á. & Trefethen, L. N. (2000) A spectral Petrov–Galerkin formulation for pipe flow. I. Linear stability and transient growth. Technical Report NA 00/18, Oxford University Computing Laboratory. [181, 182]

Miles, J. W. (1961) On the stability of heterogeneous shear flows, Part 2, *J. Fluid Mech.* **16**, 209–227. [194]

Morkovin, M. V. (1969) The many faces of transition, in *Viscous Drag*, ed. C. S. Wells (Plenum Press, New York). [211]

Nakamura, I. (1976) Steady wake behind a sphere, *Phys. Fluids* **19**, 5–8. [225]

Nakayama, Y. (ed.) (1988) *Visualized Flow* (Pergamon, Oxford). [9, 133, 174, 220, 227, 228, 230]

Natarajan, R. & Acrivos, A. (1993) The instability of the steady flow past spheres and disks, *J. Fluid Mech.* **254**, 323–344. [225]

Neu, J. C. (1984) The dynamics of stretched vortices, *J. Fluid Mech.* **143**, 253–276. [53]

Newhouse, S., Ruelle, D. & Takens, F. (1978) Occurrence of strange axiom A attractors near quasi-periodic flows on T^m, $m \geq 3$, *Commun. Math. Phys.* **64**, 35–40. [213]

Nishioka, M., Iida, S. & Ichikawa, Y. (1975) An experimental study of the stability of plane Poiseuille flow, *J. Fluid Mech.* **72**, 731–751. [176]

Nishioka, M., Asai, M. & Iida, S. (1980) An experimental investigation of secondary instability, pp. 37–46 in *Laminar-Turbulent Transition*, eds. R. Eppler & H. Fasel (Springer-Verlag, Berlin). [175]

Noether, F. (1921) Das Turbulenz problem, *Z. angew. Math. Mech.* **1**, 125–138, 218–219. [74]

Oberbeck, A. (1879) Ueber die Wärmleitung der Flüssigkeiten bei Berücksichtigung der Strömungen infolge von Temperaturdifferenzen, *Ann. Phys. Chem.* **7**, 271–292. [95]

Oertel, H. & Kirchartz, K. R. (1979) pp. 355–366 in *Recent Developments in Theoretical and Experimental Fluid Mechanics*, eds. U. Muller, K. G. Roesner & B. Schmidt (Springer-Verlag, Berlin). [109]

Onsager, L. (1949) Statistical hydrodynamics, *Nuovo Cimento* **6**(2), 279–287, suppl. ser. IX. [156]

Orr, W. M'F. (1907a) Stability or instability of the steady motions of a perfect liquid, *Proc. Roy. Irish Acad.* A **27**, 9–69. [184, 186]

Orr, W. M'F. (1907b) The stability or instability of the steady motions of a liquid, *Proc. Roy. Irish Acad.* A **27**, 69–138. [83, 157]

Orszag, S. A. & Patera, A. T. (1983) Secondary instability of wall-bounded shear flows, *J. Fluid Mech.* **128**, 347–385. [176]

Palm, E., Weber, J. E. & Kvernold, O. (1972) On steady convection in a porous medium, *J. Fluid Mech.* **54**, 153–161. [119]

Pearson, J. R. A. (1958) On convection cells induced by surface tension, *J. Fluid Mech.* **4**, 489–500. [121]

Pekeris, C. L. (1936) On the stability problem in hydrodynamics, *Proc. Camb. Phil. Soc.* **32**, 55–66. [199]

Pellew, A. & Southwell, R. V. (1940) On maintained convective motion in a fluid heated from below, *Proc. Roy. Soc. Lond. A* **176**, 312–343. [108]

Pomeau, Y. & Manneville, P. (1980) Intermittent transition to turbulence in dissipative dynamical systems, *Commun. Math. Phys.* **74**, 189–197. [213]

Prandtl, L. (1921) Bemerkungen über die Enstehung der Turbulenz, *Z. angew. Math. Mech.* **1**, 431–436. Also in *Phys. Z.* **23** (1922), 19–25, and *Gesammelte Abhandlungen* **2**, 687–696. [161]

Prandtl, L. (1935) The mechanics of viscous fluids, in *Aerodynamic Theory*, ed. W. F. Durand (Springer, Berlin) **3**, Division G, 34–208. [161]

Provansal, M., Mathis, C. & Boyer, L. (1987) Bénard–Kármán instability: transient and forced regimes, *J. Fluid Mech.* **182**, 1–22. [221, 224]

Raetz, G. S. (1959) A new theory of the cause of transition in fluid flows, *Norair Rep.* NOR-59-383, Hawthorne, CA. [177]

Rayleigh, J. W. S. (1879) On the instability of jets, *Proc. Lond. Math. Soc.* **10**, 4–13. Also in *Sci. Papers* **1**, 361–371. [62]

Rayleigh, J. W. S. (1880) On the stability, or instability, of certain fluid motions, *Proc. Lond. Math. Soc.* **11**, 57–70. Also in *Sci. Papers* **1**, 474–487. [136, 143, 144, 149, 181–183]

Rayleigh, J. W. S. (1883) Investigation of the character of the equilibrium of an incompressible heavy fluid of variable density, *Proc. Lond. Math. Soc.* **14**, 170–177. Also in *Sci. Papers* **2**, 200–207. [51, 60, 193]

Rayleigh, J. W. S. (1892) On the question of the stability of the flow of fluids, *Phil. Mag.* **34**(5), 59–70. Also in *Sci. Papers* **3**, 575–584. [168, 198]

Rayleigh, J. W. S. (1894) *The Theory of Sound*, 2nd edn. (Macmillan, London). [187]

Rayleigh, J. W. S. (1916a) On convection currents in a horizontal layer of fluid, when the higher temperature is on the under side, *Phil. Mag.* **32**(6), 529–546. Also in *Sci. Papers* **6**, 432–446. [93, 98]

Rayleigh, J. W. S. (1916b) On the dynamics of revolving fluids, *Proc. Roy. Soc. Lond. A* **93**, 148–154. Also in *Sci. Papers* **6**, 447–453. [124]

Reddy, S. C. & Henningson, D. S. (1993) Energy growth in viscous channel flows, *J. Fluid Mech.* **252**, 209–238. [177]

Reid, W. H. (1960) Inviscid modes of instability in Couette flow, *J. Math. Anal. Applics.* **1**, 411–422. [137]

Reynolds, O. (1883) An experimental investigation of the circumstances which determine whether the motion of water shall be direct or sinuous, and of the law of resistance in parallel channels, *Phil. Trans. Roy. Soc. Lond. A* **174**, 935–982. Also in *Sci. Papers* **2** (1901), 51–105. [2–4, 29, 35, 54, 74, 181]

Reynolds, O. (1895) On the dynamical theory of incompressible viscous fluids and the determination of the criterion, *Phil. Trans. Roy. Soc. Lond. A* **186**, 123–164. Also in *Sci. Papers* **2** (1901), 535–577. [83, 84]

Romanov, V. A. (1973) Stability of plane-parallel Couette flow, *Funkcional Anal. i Proložen* **7**(2), 62–73. Translated in *Functional Analysis and Its Applications* **7**, 137–146. [168]

Rossby, H. T. (1969) A study of Bénard convection with and without rotation, *J. Fluid Mech.* **36**, 309–335. [101, 103]

Ruelle, D. & Takens, F. (1971) On the nature of turbulence, *Commun. Math. Phys.* **20**, 167–192. [213]

Saffman, P. G. (1992) *Vortex Dynamics* (Cambridge University Press). [151, 154]

Saffman, P. G. & Taylor, G. I. (1958) The penetration of a fluid into a porous medium or Hele-Shaw cell containing a more viscous liquid, *Proc. Roy. Soc. Lond. A* **245**, 312–329. [61]

Salwen, H., Cotton, F. W. & Grosch, C. E. (1980) Linear stability of Poiseuille flow in a circular pipe, *J. Fluid Mech.* **98**, 273–284. [35, 181]

Saric, W. S. (1994) Görtler vortices, *Ann. Rev. Fluid Mech.* **26**, 379–409. [134]

Saric, W. S. & Thomas, A. S. W. (1984) Experiments on the subharmonic route to turbulence in boundary layers, pp. 117–22 of *Turbulence and Chaotic Phenomena in Fluids*, ed. T. Tatsumi (North-Holland, Amsterdam). [177]

Schensted, I. V. (1960) Contributions to the theory of hydrodynamic stability, Ph.D. thesis, University of Michigan. [157]

Schlichting, H. (1933) Zur Entstehung der Turbulenz bei der Plattenströmung, *Nachr. Ges. Wiss. Göttingen, Math.-phys. Klasse* 181–208. [168, 173]

Schmid, P. J. & Henningson, D. S. (2001) *Stability and Transition in Shear Flows* (Springer, New York). [9, 172, 177]

Schmid-Burgk, J. (1965) Zweidimensionale selbstkonsistente Lösungen stationären Wlassovgleichung für Zweikomponentenplasmen, Diplomarbeit, Ludwig-Maximilians-Universität, München. [190]

Schubauer, G. B. & Skramstad, H. (1947) Laminar boundary-layer oscillations and transition on a flat plate, *J. Nat. Bur. Standards* **38**, 251–292. Also in *Rep. Nat. Adv. Comm. Aero., Wash.* No. 909 (1948). [158, 173, 219]

Serrin, J. (1959) On the stability of viscous fluid motions, *Arch. Rat. Mech. Anal.* **3**, 1–13. [84, 226]

Sexl, T. (1927a) Zur Stäbilitätsfrage der Poiseuilleschen und Couetteschen Strömung, *Ann. Phys., Leipzig* **83**(4), 835–848. [179]

Sexl, T. (1927b) Über dreidimensionale Störung der Poiseuilleschen Strömung, *Ann. Phys., Leipzig* **84**(4), 807–822. [179]

Silveston, P. L. (1958) Wärmedurchgang in waagerechten Flüssigkeits-schichten, *Forsch. Gebiete Ingenieurwes.* **24**, 29–32. [101, 103]

Smith, A. M. O. & Gamberoni, N. (1956) Transition, pressure gradient, and stability theory, Douglas Aircraft Co., Inc., Rep. No. ES 26388. [177]

Smith, F. T. (1979a) On the non-parallel flow stability of the Blasius boundary layer, *Proc. Roy. Soc. Lond. A* **366**, 91–109. [166]

Smith, F. T. (1979b) Nonlinear stability of boundary layers for disturbances of various sizes, *Proc. Roy. Soc. Lond. A* **368**, 573–589. [166]

Sobey, I. J. (2000) *Introduction to Interacting Boundary Layer Theory* (Oxford University Press). [226]

Sobey, I. J. & Drazin, P. G. (1986) Bifurcation of two-dimensional channel flows, *J. Fluid Mech.* **17**, 263–287. [19, 226, 231]

Sommerfeld, A. (1908) Ein Beitrag zur hydodynamische Erklaerung der turbulenten Fluessigkeitsbewegungen, *Proc. 4th Intl Congr. Math..* Rome, vol. III, pp. 116–124. [157]

Sorokin, V. S. (1961) Nonlinear phenomena in closed flows near critical Reynolds numbers, *Prikl. Mat. Mekh.* **25**, 248–258. Translated in *J. Appl. Math. Mech.* **25**, 366–381. [91]

Squire, H. B. (1933) On the stability of three-dimensional disturbances of viscous flow between parallel walls, *Proc. Roy. Soc. Lond. A* **142**, 621–628. [141]

Srulijes, J. A. (1979) Zellularkonvektion in Behältern mit horizontalen Temperaturgradienten, Dissertation, Universität Karlsruhe. [104]

Stern, M. E. (1960) The 'salt fountain' and thermohaline convection, *Tellus* **12**, 172–175. [120]

Stokes, G. G. (1880) Supplement to a paper on the theory of oscillatory waves, *Math. Phys. Papers* **I** (1880), 314–326. [74]

Stone, P. H. (1969) The meridional structure of baroclinic waves, *J. Atmos. Sci.* **26**, 376–389. [219]

Straughan, B. (1982) *Instability, Nonexistence and Weighted Energy Methods in Fluid Dynamics and Related Theories* (Pitman, London). [84]

Stuart, J. T. (1960) On the non-linear mechanics of wave disturbances in stable and unstable parallel flows. Part 1. The basic behaviour in plane Poiseuille flow, *J. Fluid Mech.* **9**, 353–370. [75, 176]

Stuart, J. T. (1967) On finite amplitude oscillations in laminar mixing layers, *J. Fluid Mech.* **29**, 417–440. [190]

Swift, J. & Hohenberg, P. C. (1977) Hydrodynamic fluctuations at the convective instability, *Phys. Rev. A* **15**, 319–328. [116]

Synge, J. L. (1933) The stability of heterogeneous fluids, *Trans. Roy. Soc., Canada* **27**(3), 1–18. [125, 135]

Synge, J. L. (1938) Hydrodynamic stability, *Semi-centenn. Publ. Amer. Math. Soc.* **2**, 227–269. [162]

Taneda, S. (1959) Downstream development of wakes behind cylinders, *J. Phys. Soc. Japan* **14**, 843–848. [219]

Taneda, S. (1978) Visual observations of the flow past a sphere at Reynolds numbers between 10^4 and 10^6, *J. Fluid Mech.* **85**, 187–192. [225]

Tatsumi, T. & Gotoh, K. (1960) The stability of free boundary layers between two uniform streams, *J. Fluid Mech.* **7**, 433–441. [203]

Taylor, G. I. (1915) Eddy motion in the atmosphere, *Phil. Trans. Roy. Soc. Lond. A* **215**, 1–26. Also in *Sci. Papers* **2**, 1–23. [144, 161]

Taylor, G. I. (1923) Stability of a viscous liquid contained between two rotating cylinders, *Phil. Trans. Roy. Soc. Lond. A* **223**, 289–343. Also in *Sci. Papers* **4**, 34–85. [125, 127, 129, 130]

Taylor, G. I. (1931) Effect of variation of density on the stability of superposed streams of fluid, *Proc. Roy. Soc. Lond. A* **132**, 499–523. Also in *Sci. Papers* **4**, 34–85. [194]

Taylor, G. I. (1938) Some recent developments in the study of turbulence, *Proc. 5th Intern. Congr. Appl. Mech.*, 294–310 (Wiley, New York). [159]

Taylor, G. I. (1950) The instability of liquid surfaces when accelerataed in a direction perpendicular to their planes. I, *Proc. Roy. Soc. Lond. A* **201**, 192–196. Also in *Sci. Papers* **3**, 532–536. [51, 193]

Thomas, L. H. (1953) The stability of plane Poiseuille flow, *Phys. Rev.* **91**(2), 780–783. [168]

Thorpe, S. A. (1968) A method of producing a shear flow in a stratified fluid, *J. Fluid Mech.* **32**, 693–704. [56]

Thorpe, S. A. (1969) Experiments on the instability stratified shear flows: immiscible fluids, *J. Fluid Mech.* **39**, 25–48. [54]

Thorpe, S. A. (1985) Laboratory observations of secondary structures in Kelvin–Helmholtz billows and consequences for ocean mixing, *Geophys. Astrophys. Fluid Dyn.* **34**, 175–190. [55]

Threlfall, D. C. (1975) Free convection in low-temperature gaseous helium, *J. Fluid Mech.* **67**, 17–28. [103]

Tollmien, W. (1929) Über die Entstehung der Turbulenz, *Nachr. Ges. Wiss. Göttingen, Math.-phys. Klasse*, 21–44. Translated as 'The production of turbulence', *Tech. Memor. Nat. Adv. Comm. Aero., Wash.* No. 609 (1931). [158, 159, 164, 168, 173]

Tollmien, W. (1935) Ein allgemeines Kriterium der Instabilität laminarer Geschwindigkeitsverteilungen, *Nachr. Wiss. Fachgruppe, Göttingen, Math.-phys. Klasse* **1**, 79–114. Translated as 'General instability criterion of laminar velocity distributions', *Tech. Memor. Nat. Adv. Comm. Aero., Wash.* No. 792 (1936). [146, 173]

Trefethen, L. N., Trefethen, A. N., Reddy, S. C. & Driscoll, T. A. (1993) Hydrodynamic stability without eigenvalues, *Science* **261**, 578–584. [39, 206]

Tutty, O. R. (1996) Nonlinear development of flow in channels with non-parallel walls, *J. Fluid Mech.* **326**, 265–284. [232]

Van Dyke, M. (1982) *An Album of Fluid Motion* (Parabolic Press, Stanford). [4, 9, 100, 109, 127, 128, 174, 216, 217, 220, 225]

van Ingen, J. L. (1956) A suggested semi-empirical method for the calculation of the boundary layer transition region, Dept Aeronaut. Engng, Tech. Univ., Delft, Reps. Nos. VTH 71, 74. [177]

von Helmholtz, H. *See* Helmholtz, H. von.

von Kármán, T. *See* Kármán, T. von.

Waleffe, F. (1995) Hydrodynamic stability and turbulence: beyond transients to a self-sustaining process, *Studies Appl. Math.* **95**, 319–343. [39, 196]

Warn, T. & Warn, H. (1978) The evolution of a nonlinear critical level, *Studies Appl. Math.* **59**, 37–71. [33]

Watson, J. (1960a) Three-dimensional disturbances in flow between parallel planes, *Proc. Roy. Soc. Lond. A* **254**, 562–569. [198]

Watson, J. (1960b) On the non-linear mechanics of wave disturbances in stable and unstable parallel flows. Part 2. The development of a solution for plane Poiseuille flow and for plane Couette flow, *J. Fluid Mech.* **9**, 353–370. [75, 176]

Williamson, C. H. K. (1996) Vortex dynamics in the cylinder wake, *Ann. Rev. Fluid Mech.* **28**, 477–539. [223]

Willis, G. E. & Deardorff, J. W. (1967) Development of short-period temperature fluctuations in thermal convection, *Phys. Fluids* **10**, 931–937. [103]

Wilson, S. (1969) The development of plane Poiseuille flow, *J. Fluid Mech.* **38**, 793–806. [201]

Young, A. D. (1989) *Boundary Layers* (BSP Professional Books, Oxford). [173]

Yudovich, V. I. (1989) *The Linearization Method in Hydrodynamic Stability Theory*, Translations Math. Monographs, vol. 74 (American Mathematical Society, Providence, RI) [Originally published in Russian in 1984.] [38, 184, 186]

Motion Picture Index

In the 1960s the Education Development Center made several films for the National Committee for Fluid Mechanics Films. It made 16 mm sound films, each one taking about 30 minutes to project, and 8 mm film loops, each one typically taking about 4 minutes. Most of the loops contain extracts from the long films. Many of the films are of demonstrations and experiments of hydrodynamic instability, and most of these are still excellent. These films were subsequently converted into videos, and the videos are still for sale. They are distributed in the United States and many other countries by the Encyclopaedia Britannica Educational Corporation, 310 South Michigan Avenue, Chicago, Illinois 60604, and elsewhere by some special distributors.

The first compact disk with motion pictures of fluid mechanics was published in 2000 by Cambridge University Press. It reproduces some short extracts from the Education Development Center films, and many other films and videos of laboratory experiments and natural phenomena. It also has short videos of computer simulations of various flows. Most of these videos are excellent, and many are of hydrodynamic instability. Many of these motion pictures have been cited in the text, and their index is below. The initials CD have been used for the compact disk, F for a film, and FL for a film loop.

Brown, F. N. M. (FL1964) *Stages of boundary-layer instability and transition*. B/W–No. S–FM092. [174]

Bryson, A. E. (F1967) *Waves in fluids*. B/W–No. 21611. [57]

Bryson, A. E. (FL1967) *Small-amplitude gravity waves in an open channel*. B/W–No. S–FM139. [57]

Coles, D. (FL1963a) *Instabilities in circular Couette flow*. B/W–No. S–FM031. [127]

Coles, D. (FL1963b) *Examples of turbulent flow between concentric rotating cylinders*. B/W–No. S–FM032. [127]

Homsy, G. M., Aref, H., Breuer, K. S., Hochgreb, S., Koseff, J. R., Munson, B. R., Powell, K. G., Robertson, C. R. & Thoroddsen, S. T. (CD2000) *Multi-media Fluid Mechanics* (Cambridge University Press). [6, 9, 127, 174, 216, 220, 225]

Lippisch, A. M. (FL1964) *Tollmien–Schlichting waves*. B/W–No. S–FM023. [174]

Long, R. R. (F1968) *Stratified flow*. Color–No. 21618. [57]

Mollo-Christensen, E. L. (F1968) *Flow instabilities*. Color–No. 21619. [59]

Mollo-Christensen, E. L. & Wille, R. (FL1968) *Experimental study of a flow instability*. B/W–No. S–FM147. [57]

Stewart, R. W. (F1968) *Turbulence*. Color–No. 21626. [6]

Stewart, R. W. (FL1968) *Laminar and turbulent pipe flow*. Color–No. S–FM134. [6]

Trefethen, L. M. (F1965) *Surface tension in fluid mechanics*. Color–No. 21610. [65]

Trefethen, L. M. (FL1965) *Breakup of liquid into drops*. Color–No. S–FM076. [65]

Index

Printed in the United States
By Bookmasters